AQA Geography

A LEVEL & AS

Exam Practice and Skills

Bob Digby   Tim Bayliss

OXFORD
UNIVERSITY PRESS

Great Clarendon Street, Oxford, OX2 6DP, United Kingdom

Oxford University Press is a department of the University of Oxford.
It furthers the University's objective of excellence in research,
scholarship, and education by publishing worldwide. Oxford is a
registered trade mark of Oxford University Press in the UK and in
certain other countries

British Library Cataloguing in Publication Data
Data available

978-0-19-843258 6

10 9 8 7

Paper used in the production of this book is a natural, recyclable
product made from wood grown in sustainable forests.
The manufacturing process conforms to the environmental
regulations of the country of origin.

Printed and bound by CPI Group UK Ltd, Croydon, CR0 4YY

Acknowledgements

The publisher and authors would like to thank the following for
permission to use photographs and other copyright material:

Cover: wanderluster/iStockphoto; Cover: Volodymyr Nikulishyn/
Shutterstock. **p9**: Bob Digby; **p21**: Bob Digby; **p118**: Phil Judd/
Cartoonstock; **p134, 176**: Ordnance Survey © Crown copyright
and Database rights **2019**.; **p136-8, 152, 169**: Tim Bayliss; **p169**:
Shutterstock; **p172**: Les Gibbon/Alamy Stock Photo.

Artwork by Aptara Inc., Kamae Design, Lovell Johns, Barking Dog
Art, and Q2A Media Services Inc.

Every effort has been made to contact copyright holders of material
reproduced in this book. Any omissions will be rectified in
subsequent printings if notice is given to the publisher.

Contents

Exam Practice

1	Introduction	4
1.1	Preparing for success	4
1.2	The exams – topics and question types	5
1.3	Know your command words	6
1.4	Handy hints for exams	7
1.5	Understanding assessment objectives	8
1.6	Managing the 9-mark questions	10
1.7	Managing the 20-mark questions	12

2	On your marks	14
2.1	Dealing with 6-mark questions in the exam	14
2.2	6-mark questions using 'Analyse'	15
2.3	6-mark questions using 'Assess'	21
2.4	9-mark questions on Hazards	27
2.5	9-mark questions on Ecosystems under stress	34
2.6	9-mark questions on Contemporary urban environments	41
2.7	9-mark questions on Population and the environment	48
2.8	9-mark questions on Resource security	55
2.9	20-mark questions on the Carbon cycle	62
2.10	20-mark questions on Hot desert systems and landscapes	71
2.11	20-mark questions on Coastal systems and landscapes	80
2.12	20-mark questions on Glacial systems and landscapes	89
2.13	20-mark questions on Global systems and global governance	98
2.14	20-mark questions on Changing places	108

Geographical skills

Introduction		118

3	Cartographic Skills	120
3.1	Choropleth maps, dot maps and proportional circles	120
3.2	Isolines on maps and diagrams	124
3.3	Flow, desire and trip lines	128
3.4	Ordnance Survey maps	132

4	Graphical skills	136
4.1	Sketching, labelling and annotation	136
4.2	Graphs	140
4.3	Scattergraphs	144
4.4	Triangular graphs	148
4.5	Radial charts	152
4.6	Additional higher order graphical skills	156

5	Statistical skills	160
5.1	Measures of central tendency	160
5.2	Variance and standard deviation	164
5.3	The Spearman rank correlation coefficient test	168
5.4	The Chi-squared test	172

Ordnance Survey map symbols	176

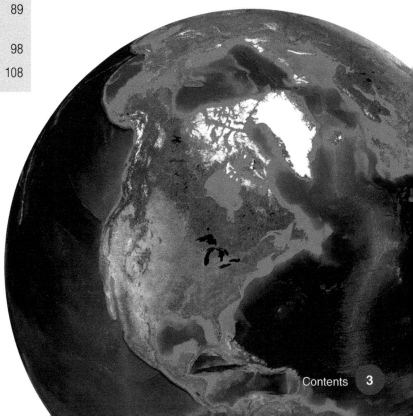

Introduction
1.1 Preparing for success

To be successful in your exams, revision is essential. But so too is exam preparation – knowing what each paper contains, and the sorts of questions that examiners can ask. That's why this book has been written. It contains key guidance that you need to answer exam questions for AQA's A Level Geography specification.

How to use this book

This book is one of five OUP publications to support your learning. The others are:

- *AQA Geography A Level and AS Physical Geography* student book.
  This covers topics for Paper 1 in the specification.
- A separate revision guide covering content in the physical geography student book.
- *AQA Geography A Level and AS Human Geography* student book.
  This covers topics for Paper 2 in the specification.
- A separate revision guide covering content in the human geography student book.

An introduction (pages 4–13)

This section contains details about:

- the exam papers you'll be taking
- topics you need to revise for each exam paper
- question styles and command words
- understanding assessment objectives
- preparing for longer answers carrying 9 and 20 marks.

On your marks (pages 14–117)

This section contains detailed guidance about how to answer questions using extended writing for 6, 9 and 20 marks, focusing on Papers 1 and 2.

- Those sections for 6- and 9-mark questions contains space for you to write practice exam answers, with opportunities to assess exam answers (including your own!) so that you get to know how to write good quality, focused responses.
- Those sections covering 20-mark questions will give you the opportunity to write your own exam answers, but these will need to be completed on file paper.
- However, each of the 20-mark sections will contain two written answers for you to read and assess, in addition to your own answer.

Geographical skills (pages 118–176)

This section contains detailed support for geographical skills, particularly those which you will find useful for:

- **learning topics** where particular skills enhance your understanding of a theme (e.g. the Gini Index and Global governance)
- **your Non-Examined Assessment (NEA)**, because each skill is presented clearly in a way for you to use when you are analysing data
- **revision**, because some skills will be assessed in exam Papers 1 and 2.

In this section you'll learn about the two exam papers.

AQA's A Level Geography specification has 11 topics (some of which are optional), assessed in two exams, Paper 1 (physical topics) and Paper 2 (human topics).

In **Paper 1**, you will study **three** physical geography topics from the six in the specification:

- *Water and carbon cycles*, which is compulsory.
- One of **either** *Hot desert systems and landscapes*, **or** *Coastal systems and landscapes*, **or** *Glacial systems and landscapes*.
- One of **either** *Hazards* **or** *Ecosystems under stress*.

Paper 1 is 2 hours and 30 minutes in length, and counts for 40% of your final mark.

In **Paper 2**, you will study **three** human geography topics from the five in the specification:

- *Global systems and global governance*, which is compulsory.
- *Changing places*, which is also compulsory.
- One of **either** *Contemporary urban environments* **or** *Population and the environment* **or** *Resource security*.

Paper 2 is also 2 hours and 30 minutes in length, and counts for 40% of your final mark.

Paper 1

Paper 1 has **three** sections, each assessing physical topics.
Figure **1** summarises each section.

Section	Marks	Topics	Details of questions
A	36	Question 1 *Water and carbon cycles*	• 1 × 4-mark short paragraph • 2 × 6-mark paragraphs • 1 × 20-mark essay
B	36	**Either** question 2 *Hot desert systems and landscapes*, **or** question 3 *Coastal systems and landscapes*, **or** question 4 *Glacial systems and landscapes*	Each question has: • 1 × 4-mark short paragraph • 2 × 6-mark paragraphs • 1 × 20-mark essay
C	48	**Either** question 5 *Hazards* **or** question 6 *Ecosystems under stress*	Each question has: • 4 × 1-mark multiple choice questions • 1 × 6-mark paragraph • 2 × 9-mark longer paragraphs • 1 × 20-mark essay

🔼 **Figure 1** *Paper 1 topics and question styles*

Paper 2

Paper 2 also has **three** sections, each assessing human topics.
Figure **2** summarises each section.

Section	Marks	Topics	Details of questions
A	36	Question 1 *Global systems and global governance*	• 1 × 4-mark short paragraph • 2 × 6-mark paragraphs • 1 × 20-mark essay
B	36	Question 2 *Changing places*	• 1 × 4-mark short paragraph • 2 × 6-mark paragraphs • 1 × 20-mark essay
C	48	**Either** question 3 *Contemporary urban environments* **or** question 4 *Population and the environment* **or** question 5 *Resource security*	Each question has: • 4 × 1-mark multiple choice questions • 1 × 6-mark paragraph • 2 × 9-mark longer paragraphs • 1 × 20-mark essay

🔼 **Figure 2** *Paper 2 topics and question styles*

In this section you'll learn about command words.

Command words

AQA's A Level Geography examiners use the command words shown in Figure **1**. Some are more challenging than others. There is no rule concerning the number of marks attached to each command word, but clearly questions asking you to 'explain' are likely to be worth fewer marks than those asking you to 'evaluate'.

In Figure **3**, low-tariff questions, i.e. those carrying fewer marks, use the command words 1–4 and high-tariff questions use the command words 5–8.

Command word	Definition
1 Suggest	For an unseen resource (e.g. photo, graph or text), provide a well-reasoned explanation of how or why something may occur, without necessarily knowing the answer. This assesses AO2.
2 Explain	Give a reasoned explanation of how or why something occurs. An explanation requires a justification/exemplification of a point. This assesses AO1.
3 Examine	Consider a range of reasons for something and, at the same time, pick out one or two reasons which best explain why something is happening. This assesses AO1 (what you know) and AO2 (judgements made).
4 Analyse	Break something down into individual components/processes and say how each one contributes to the question's theme/topic and how components/processes work together. This assesses AO2.
5 Assess	Use evidence to determine the relative significance of something. Give balanced consideration to all factors and identify which are the most important. This assesses AO2 as it is all about making a judgement.
6 Evaluate	Consider several factors, ideas or arguments; or weigh up the value or success of something, and give an evidenced and balanced judgement or conclusion. This assesses AO2 in the way you construct an argument.
7 To what extent?/ How far do you agree ...?	Form and express a view about the validity of a view or statement, by examining the evidence available or considering the merits of different sides of an argument. This assesses AO2 in the way you construct an argument.
8 Justify	Give reasons for a viewpoint or idea, or reasons why particular actions should be taken or choices made. This assesses AO2 in the way you construct an argument.

Figure 3 Command words used in AQA's A Level Geography exams

There are a few variations:

- **Analyse** is used in 6- and 9-mark questions. 'On your marks' sections 2.2 and 2.8 have examples of questions using this command word.
- **Assess** is used in anything from 6- through to 20-mark questions. 'On your marks' sections 2.3, 2.4 and 2.9 have examples of questions using this command word.
- **Evaluate** is also used in anything from 6- through to 20-mark questions. 'On Your Marks' sections 2.6, 2.10, 2.11 and 2.12 have examples of questions using this command word.

In this section you'll learn a few handy hints about coping with exams.

Understanding marks

Examiners use two types of mark scheme; point marked for shorter answers (up to 4 marks) and level marked for extended written answers (6 or more marks).

Shorter questions (worth 1–4 marks)

Questions carrying 1–4 marks are **point marked**. For every correct point that you make, you earn a mark.

- These may be 1-mark questions, e.g. the multiple choice items in each option topic (Paper 1: *Hazards* and *Ecosystems under threat*; Paper 2: *Contemporary urban environments*, *Population and the environment*, and *Resource security*).
- Other than these, the only point-marked questions occur as 4-mark questions in each of the core topics (*Water and carbon cycles*, *Hot desert/Coastal/Glacial systems and landscapes*, *Global governance and globalisation*, and *Changing places*).

Extended questions (worth 6–20 marks)

Questions carrying 6 or more marks are **level marked**, and are far more common at A Level. The examiner reads the whole answer, then uses a set of criteria – known as **levels** – to judge its qualities.

- There are **three** levels for questions carrying 6 or 9 marks, and **four** levels for questions of 20 marks.
- In each case, the *highest level earns most* marks.

In this book:

- sections 2.1–2.3 will prepare you for 6-mark questions
- sections 2.4–2.8 will prepare you for 9-mark questions
- sections 2.9–2.14 will prepare you for 20-mark questions.

You can also read general guidance about answering 9-mark questions in section 1.6, and about 20-mark questions in section 1.7.

Staying on track – getting the timing right

Timing is one of the biggest factors affecting marks and final grades. Those candidates who complete questions in the time available always do better than those who do not.

- Remember each exam paper has 150 minutes in which you have to tackle 120 marks. That's 4 marks every 5 minutes.
- Use this to calculate how long you should spend on each question – 45 minutes on 36-mark questions and an hour on 48-mark questions.
- Within each question, keep track of time; 5 minutes for 4 marks, 7–8 minutes for 6 marks, 11–12 minutes for 9 marks, and 25 minutes for 20-mark questions.

In this section you'll learn about assessment objectives

What are the assessment objectives?

An assessment objective is the key tool used by examiners to decide what you should know, understand and be able to do after studying Geography A Level for two years. Read this section to get a clear idea of why it will help you to know and understand how you're being assessed, as well as the topics you have to learn.

There are three assessment objectives in Geography A Level (Figure **4**):

- **Assessment objective 1** (**AO1**) – which is about your **knowledge and understanding**. It might be your knowledge and understanding of places, processes, or issues. It's basically the content of the two student textbooks in this series!
- **Assessment objective 2** (**AO2**) – which is about the way in which you **interpret and apply** your understanding to situations. For example, you could be asked to consider an argument. Imagine a question such as 'Can child labour ever be justified?'. You'd have to **weigh up evidence** (i.e. your knowledge and understanding – AO1) and frame it together into an **argument**. You might find that there are points you could make which could support the use of child labour in certain circumstances, followed by others against its use. By the end you might be able to **make a judgement** – perhaps in favour, perhaps against.

    This process of using information to develop an argument, is what AO2 is all about. Most high-mark exam questions that you'll answer will have high marks allocated to AO2 – it's an essential A Level skill.
- **Assessment objective 3** (**AO3**) – which is about using geographical skills in formulating questions, in thinking about methods of data collection, of manipulating data, presenting them, and drawing conclusions. You should recognise that this is exactly what you've been doing in writing up your individual investigation (known as the Non-Examined Assessment, or NEA). But you will have to use those same skills in some situations in the exam as well.

AO1 questions use command words such as **Explain**. You also need knowledge and understanding for exam questions assessing AO2.

AO2 questions use command words such as **Suggest, Assess** or **Evaluate**, using your knowledge and understanding (AO1).

AO3 questions use the command word **Calculate** (for stats questions) or **Analyse** (for questions about data interpretation).

	Objective	%	Marks
AO1	Demonstrate knowledge and understanding of places, environments, concepts, processes, interactions and change, at a variety of scales.	30–40%	90–120
AO2	Apply knowledge and understanding in different contexts to interpret, analyse and evaluate geographical information and issues.	30–40%	90–120
AO3	Use a variety of relevant quantitative, qualitative and fieldwork skills to: • investigate geographical questions and issues • interpret, analyse and evaluate data and evidence • construct arguments and draw conclusions.	20–30%	60–90

Figure 4 *Assessment objectives used in AQA's A Level Geography examinations*

Marks allocated to assessment objectives

Figure **5** shows which AOs are being assessed in each paper.

- Papers 1 and 2 largely assess **AO1** (knowledge and understanding) and **AO2** (application). This means that you really need to know your material, because **AO1** counts for 54 of the 120 marks on each paper.
- You also need to know how to argue a case, because **AO2** counts for 48 of the 120 marks on each paper.
- Similarly, some marks will assess **AO3**. That's because a few questions will assess your ability to read, manipulate, and interpret a resource, e.g. a map, diagram or data.

	AO1 marks	AO2 marks	AO3 marks	Total
Paper 1	54	48	18	**120**
Paper 2	54	48	18	**120**
NEA	0	6	54	**60**
Total	**108**	**102**	**90**	**350**

Figure 5 *Marks used for each assessment objective in AQA's A Level Geography exams, as shown in the Sample Assessment Materials (SAMs)*

Marks attached to each command word

Section 1.3 shows that some command words are more challenging than others. Figure **6** shows this in more detail. Notice also that 'Explain' only assesses one assessment objective – AO1; you need to know the answer to be able to explain it. However, most command words in Figure **6** assess more than one assessment objective.

- For example, in Papers 1 and 2, you may be given a resource, like Figure **7** below, about which you are asked a 6–mark question, such as, '*Using Figure **7** and your own knowledge, assess which geomorphic processes have been most influential in the cliff profile'*.
- In this case, you need to recognise the cliff and the processes that affect it – so 2 of the 6 marks are for AO1. The remaining marks are for applying what you know. Only some coastal processes (i.e. sub-aerial processes) have affected the cliff and you need to select which ones – that's AO2.

As marks increase, so AO2 becomes more important.

- Figure **6** shows that 20-mark questions using 'Assess' are split – 10 marks for AO1, and 10 for AO2.
- A question such as '*Assess the extent to which urban regeneration depends upon rebranding for its success*' would therefore carry 10 marks for knowledge and understanding, and 10 for AO2. You can see how examiners mark responses like this in section 1.7.

Command word	Total Mark	AO1	AO2	AO3
Explain	**4**	4		
Analyse (using a resource)	**6**			6
Analyse (long paragraph)	**9**	4	5	
Assess (using a resource)	**6**			6
Assess (paragraph)	**6**	2	4	
Assess (long paragraph)	**9**	4	5	
Assess (extended essay)	**20**	10	10	
To what extent does/How far do you agree that …?	**9**	4	5	
To what extent does/How far do you agree that …?	**20**	10	10	
Evaluate (using a resource)	**6**			6
Evaluate (long paragraph)	**9**	4	5	

▲ **Figure 6** *The balance of assessment objectives in questions using particular command words*

Figure **7** could be used as a resource on which examiners would test your knowledge (AO1) and whether you can apply that knowledge to what is happening in the photo (AO2).

◄ **Figure 7** *A collapsing cliff in Cornwall*

In this section you'll learn about the format of the 9-mark questions.

Where the 9-mark questions appear

You have to answer four 9-mark questions in total.

- For Paper 1, you'll study either *Hazards* or *Ecosystems under stress*. Each has two 9-mark questions that you must answer.
- For Paper 2, you'll study either *Contemporary urban environments* or *Population and the environment* or *Resource security*. Each has two 9-mark questions that you must answer.

The command words that you'll come across will most likely be selected from the following:

- Analyse
- Assess
- Evaluate
- To what extent …? or How far do you agree …?

Understanding the assessment objectives

To tackle these questions successfully, you need to know what's required. To do that, it's important to know the assessment objectives (AOs) for 9-mark questions. They all assess AO1 and AO2.

All questions will ask some kind of discussion point – it could be a contentious viewpoint, for example, which you'll be asked to agree or disagree with. Or you could be asked to assess the importance of something such as, for example, the level of development of a country affected by hazards.

To answer questions like these, you need to be able to do two things.

- Demonstrate **knowledge and understanding** of the topic. This is **AO1** and carries 4 of the 9 marks. This is about quoting brief examples or data to support an argument, not lengthy case studies.
- **Develop an argument** and **use evidence to support it**. This is **AO2** and carries 5 of the 9 marks. For example, if you are asked, '*How far do you agree with the view that tectonic hazard events always have greater impacts on developing countries than on developed countries?*', then you need to know your evidence (that's AO1), put a case for each side of the argument by *applying* your evidence (that's AO2) and reach a conclusion *on the basis of evidence* (that's also AO2).

Getting to know the mark scheme

It's worth getting to know the mark scheme because it is common to all 9-mark questions.

- 9-mark questions are marked using levels.
- The overall quality of your answer is judged, rather than specific points.

Markers use criteria to judge each answer. The criteria are set out in Figure **8**, and can be summarised as follows:

- Accurate geographical knowledge and understanding, e.g. knowing and quoting examples (this is AO1).
- Applying your knowledge and understanding to questions you're asked. For example, in *Hazards*, you could get a question about how useful The Park model of human response to hazards can be in hazard management. In that case you'd need to know and understand what the model is (that's AO1), and consider how useful it is (that's AO2).
- Applying geographical information and ideas to interpret evidence. For example, in *Ecosystems under stress*, you could get a question asking you to assess the influence of climate change on ecosystems, so you need to understand concepts such as changing climate over time, and how evidence can be used to support or question it (that's also AO2).
- Applying geographical information and ideas to make judgements, so that your arguments are balanced. For example, in *Resource security*, you could get a question asking you to assess the environmental impacts of a major water supply scheme – so you need to weigh up what you know (i.e. the evidence) to assess and judge each impact (that's also AO2).

Level	Marks	Descriptor
3	7–9	• There is evidence of detailed knowledge, of clear understanding of concepts, and of physical and human processes. Evidence of these can be found throughout the answer. (AO1) • Knowledge and understanding is well applied to the question. (AO2) • Different concepts or links or relationships between different parts of the answer are relevant, and are expanded upon fully. (AO2) • There is evidence of well-considered analysis and evaluation, again supported with detailed evidence. (AO2) • A coherent and logical argument is sustained throughout the answer. (AO2)
2	4–6	• There is evidence of clear and relevant knowledge, of a sound understanding of concepts, and grasp of physical and human processes. Evidence of these is sustained though perhaps with minor inaccuracy. (AO1) • Knowledge and understanding is generally applied to the question. (AO2) • Different concepts or links or relationships between different parts of the answer are generally relevant, and are usually expanded upon. (AO2) • There is evidence of general analysis and evaluation, again supported with some evidence. (AO2) • An argument is presented, which is generally coherent and logical. (AO2)
1	1–3	• There is evidence of basic knowledge, and of an incomplete understanding of concepts or a grasp of physical and human processes. Evidence of these is limited and may be inaccurate. (AO1) • Knowledge and understanding is applied in very limited ways. (AO2) • Different concepts or links or relationships between different parts of the answer are basic, with limited development or expansion. (AO2) • There is evidence of basic analysis and evaluation, supported with limited evidence. (AO2) • An argument is presented, which is generally basic and rarely clear. (AO2)
0	0	• Nothing worthy of credit.

▲ **Figure 8** *Marking criteria for all 9-mark questions*

In this section you'll learn about the format of the 20-mark questions.

Where the 20-mark questions appear

You have to answer six 20-mark questions in total.

There are three 20-mark questions in Paper 1:

- one on *Water and carbon cycles*
- one on **either** *Hot desert systems and landscapes*, **or** *Coastal systems and landscapes*, **or** *Glacial systems and landscapes*
- one on **either** *Hazards* **or** *Ecosystems under stress*.

There are three more 20-mark questions in Paper 2:

- one on *Global systems and global governance*
- one on *Changing places*
- one on **either** *Contemporary urban environments* **or** *Population and the environment* **or** *Resource security*.

In this book, you'll learn how to tackle 20-mark questions in the following sections:

- *Water and carbon cycles* (2.9)
- *Hot desert systems and landscapes* (2.10)
- *Coastal systems and landscapes* (2.11)
- *Glacial systems and landscapes* (2.12)
- *Global systems and global governance* (2.13)
- *Changing places* (2.14)

Understanding the assessment objectives (AOs)

To tackle these questions successfully, you need to know what's required. To do that, it's important to know the assessment objectives (AOs) for 9-mark questions.

Like the 9-mark questions (see section 1.5), they assess two AOs. All questions require the development of an argument. To do that successfully, you must do two things:

- Demonstrate **knowledge and understanding**. This is **AO1** and carries 10 marks. This is about quoting examples or data to support an argument; you don't need to include lengthy case studies.
- **Develop an argument** and **use evidence to support it**. This is **AO2** and also carries 10 marks. For example, if you are asked to 'Assess the extent to which future demands for energy can only be met with renewable resources.', then you need to argue whether that is true or not, and reach an evidenced conclusion.

Getting to know the mark scheme

It's worth getting to know the mark scheme because it is common to all 20-mark questions.

- 20-mark questions are marked using levels – though this time using **four** levels, not three (as with 6- and 9-mark questions)
- The overall quality of your answer is judged, rather than specific points.

Remember!

Irrespective of the command word, all 20-mark questions need to have:
- a brief (1–2 sentence) introduction
- a final short paragraph conclusion.

Markers use criteria to judge each answer. The criteria are set out in Figure **9**, and can be summarised as follows:

- Accurate geographical knowledge and understanding, e.g. knowing processes, quoting examples. This is AO1.
- Applying geographical information and ideas. For example, section 2.11 explores the impacts of sea level change on coastal landscapes, so you need to consider how significant these changes are. This is AO2.
- Interpreting evidence. For example, section 2.9 asks how far global demand for energy is the most important factor modifying the carbon cycle – so you need to weigh up the evidence for and against this. This is also AO2.
- Making judgements, so that your arguments are balanced. For example, section 2.14 is about evaluating government policies in changing places – so you need to decide how far governments can actually affect change. This is also AO2.

Level	Marks	Descriptor
4	16–20	For AO1, the answer shows knowledge and understanding which: • is detailed, thorough and relevant to the argument • is accurate in explaining key concepts and processes • shows careful awareness of geographical scale and change over time (where appropriate). For AO2, the answer shows application which: • is thorough in a range of contexts • shows detailed, coherent and relevant analysis • leads to a detailed evaluative conclusion (as well as ongoing evaluation throughout the answer) which is rational and firmly based on evidence.
3	11–15	For AO1, the answer shows knowledge and understanding which is: • clear and mostly relevant to the argument • mostly clear and accurate in explaining concepts and processes • mostly aware of geographical scale and change over time. For AO2, the answer shows application which: • is sound in different contexts • shows thorough, clear and relevant analysis • leads to a clear, evaluative conclusion (with some ongoing evaluation throughout the answer) which is based on evidence.
2	6–10	For AO1, the answer shows knowledge and understanding which: • is partially relevant, with a few inaccuracies or omissions • is generally clear in explaining concepts and processes • shows some awareness of geographical scale and change over time, with a few inaccuracies. For AO2, the answer shows application which: • shows some application to the question • shows partially relevant analysis • leads to a generally sound, evaluative conclusion (though with little ongoing evaluation throughout the answer) which is sometimes based on evidence.
1	1–5	For AO1, the answer shows knowledge and understanding which: • is limited in relevance, with inaccuracies and/or omissions • shows isolated understanding of concepts and processes • shows limited awareness of geographical scale and change over time, with a number of inaccuracies. For AO2, the answer shows application which: • shows limited application to the question • lacks clarity in analysis and evaluation • leads to a limited and/or unsupported evaluative conclusion which is only loosely based on evidence.
0	0	• Nothing worthy of credit.

🔺 **Figure 9** *Marking criteria for all 20-mark questions*

In this section you'll learn how to...

- tackle 6-mark questions.

Getting to know 6-mark questions

There are two types of 6-mark question in Papers 1 and 2.

1. Questions using the command word 'Analyse' (see section 2.2).
   - They use tables of data, graphs or statistical diagrams.
   - They assess AO3 (geographical skills).
   - They appear as the second question in each core topic.

2. Questions using the command word 'Assess' (see section 2.3).
   - They use a resource stimulus (e.g. a photo), the purpose of which is to jog your thoughts.
   - They assess AO1 (knowledge and understanding) for 2 marks and AO2 (application) for 4 marks.
   - They appear as the third question in each core topic.

Questions using 'Analyse'

You'll find 6-mark questions using 'Analyse' with data resources in:

- all topics in Paper 1
- all topics in Paper 2.

The question will be in the style of '*Analyse the resources in Figures **X** and **Y***'.

How to analyse

Analyse means to dissect and make sense of – breaking down data to aid understanding. The data are likely to be a table of statistics, a graph or even a map containing data. You'll use quantitative techniques to interpret the data.

Three processes are involved:

- **Describe** what the data show.
- **Break down the data** – including the highest and lowest values, the mean, median, mode, range, dividing into quartiles, percentage change, trends, and anomalies.
- **Explain** – giving intelligent reasons for patterns and trends that you see.

Questions using 'Assess'

You'll come across 6-mark questions using 'Assess' in:

- Paper 1: *Water and carbon cycles*; *Hot desert* **or** *Coastal* **or** *Glacial systems and landscapes*.
- Paper 2: *Global systems and global governance*; *Changing places*.
- Most **optional topics** on Papers 1 and 2.

How to assess

There'll be a resource stimulus (e.g. a photo), with question wording in the style of '*Using Figure 2 and your own knowledge, assess the ...*'.

The question wants you to show what you know, but also to 'assess' how important it is.

For example, if asked to assess the processes which have led to a particular landform, you would need to say which are the most important processes, of all those that you know.

In this section you'll learn how to...

- maximise marks on 6-mark questions that use the command word 'Analyse'.

What this section is about

This section helps you to understand what is required in answering 6-mark questions that use the command word 'Analyse'.

- These questions use tables of data, graphs or statistical diagrams.
- They assess AO3 (geographical skills, including statistical skills)
- The skills you need to answer these questions involve analysing quantitative data.

Handy Hint!

You could come across 6-mark questions assessing AO3 in every section of each exam paper.

Question

Analyse the data in Figure 1. (6 marks)

	1973	1983	1993	2003	2013
	Value in US$ billions				
World	579	1838	3684	7380	18301
	Share as a % of global value				
North America	17.3	16.8	18	15.8	13.2
South & Central America	4.3	4.5	3	3	4
Europe	50.9	43.5	45.3	45.9	36.3
CIS	No data	No data	1.5	2.6	4.3
Africa	4.8	4.5	2.5	2.4	3.3
Asia and Oceania	14.9	19.1	26	26.1	31.5

Figure 1 World exports by global region 1973–2013. Note that percentage shares do not add to 100%.

Five steps to success!

Five steps to help you write top quality answers

The following five steps are used in this book to help you get the best marks.

1. **Plan your answer** – decide what to include and how to structure your answer.

2. **Write your answer** – use the answer spaces to complete your answer.

3. **Mark your answer** – use the mark scheme to self- or peer-mark your answer. You can also use this to assess sample answers in step 4 below.

4. **Sample answers** – sample answers are given to show you how to maximise marks for a question.

5. **Marked sample answer** – this is the same answer that you used for step 4 above, but is marked and annotated, so that you can compare with your own answers.

This question assesses AO3 (geographical skills) for all 6 marks.

① Plan your answer

Before attempting to answer the question, remember to **BUG** it. That means:

✓ **Box** the command word.
✓ **Underline** the following:
  - the theme
  - the focus
  - any evidence required
  - the number of points needed.
✓ **Glance** back over the question – to make sure you include everything in your answer.

Use the **BUG** on the next page to plan your own answer.

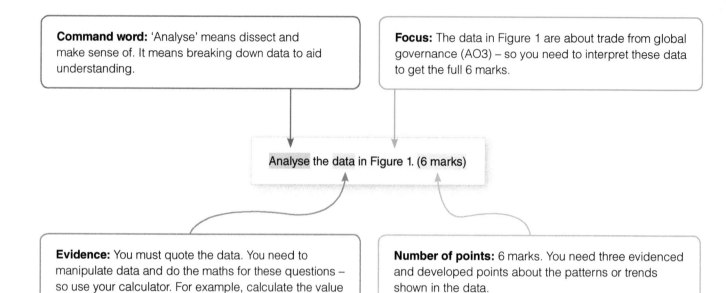

> **Command word:** 'Analyse' means dissect and make sense of. It means breaking down data to aid understanding.

> **Focus:** The data in Figure 1 are about trade from global governance (AO3) – so you need to interpret these data to get the full 6 marks.

Analyse the data in Figure 1. (6 marks)

> **Evidence:** You must quote the data. You need to manipulate data and do the maths for these questions – so use your calculator. For example, calculate the value in US$ billions worth of exports for each continent from the percentage shares of the global total.

> **Number of points:** 6 marks. You need three evidenced and developed points about the patterns or trends shown in the data.

PEEL your answer

Use PEEL notes to help structure your answer. This will help you explain your ideas to the examiner most clearly. PEEL has four stages:

- **P**oint – Make three developed points for 6 marks about the trends or changes you identify in Figure 1. Use sentences, not bullet points.
- **E**vidence – Include details from the table of data to illustrate the points or trends you identify.
- **E**xplain – Give a reason for each point you make. Use starter phrases such as: '*These changes have happened because* ...', or '*One reason is* ...'. This is part of the answer where you explain why some of the data have changed in the ways they have.
- **L**ink back to the question – Remember that 'analyse' means to pick the data apart. Don't just say '*The figures for Asia and Oceania have increased*'. Look carefully – have they increased faster at some times than others? How do the increases compare with the global pattern – is it faster or slower?

Remember the AOs!

Remember how important it is to demonstrate detailed knowledge and understanding.
- AO3 in this question is about demonstrating skills in interpreting the data.
- You need to be able to identify trends (both increases and decreases), and be able to calculate the value in US$ billions of exports for each continent from the percentage shares of the global total.

 Tip

Remember, quality not quantity

You will not be marked on the **number** of points you make, but on the **quality** of your answer. That means the quality of the content, the evidence you give, and how you link back to the question.

2 Write your answer

Analyse the data in Figure 1. (6 marks)

_____

_____

_____

_____

_____

_____

_____

_____

_____

_____

_____

_____

_____

3 Mark your answer

1. To help you to identify if the answer includes well-structured points, first highlight the:

- points in red
- explanations in orange
- evidence in blue.

2. Use the mark scheme on the next page to decide what mark to give.

- Six-mark questions are marked using **two** levels, not three (as you'll find with 9- and 20-mark questions).
- These questions are not marked using individual points, but instead you should choose a level and a mark based upon the quality of an answer as a whole.
- Remember, a top-level answer must include elements of **both** bullet points from Level 2 of the mark scheme on page 18, i.e. analysis and connections.

Level	Marks	Descriptor	Examples
2	4–6	• Clear analysis of the quantitative evidence provided which makes appropriate use of data to support. • Clear connections between different aspects of the data.	• Between 1973 and 2013, the value of global trade increased by 31.6 times. • There has been a global shift in export share. Asia and Oceania increased their share from 14.9% to 31.5%, while North America declined from 17.3% to 13.2%.
1	1–3	• Basic analysis of the quantitative evidence provided, which makes limited use of data to support. • Basic or limited connections between different aspects of the data.	• Between 1973 and 2013, the value of global exports increased from $579 billion to $18301 billion. • The share of Oceania and Asia increased while that of North America declined.
0	0	Nothing worthy of credit.	

Things to watch out for

Study Sample Answer 1 below.

- The candidate has done some things well here. Read the answer and identify which qualities of the answer are good.
- Look at the mark scheme above and work out what is missing, and what prevents the candidate from getting a top Level 2 mark.

Clues:
Look in particular at how much the candidate demonstrates:
a) analysis of data – what specific detail is there? Does the candidate simply quote simple points from the data, or has there been some manipulation of data, and identification of trends.
b) connections – does the candidate make connections between different aspects of the data?

Sample Answer 1

The value and share of world exports has seen many changes between 1973 and 2013. The value of world exports has increased by many times but the shares have changed. North America has decreased its share from 17.3% to 13.2% in 2013 (though it increased between 1973 and 1993). Meanwhile, the European share has decreased more sharply from 50.9% to 36.3%. South and Central America has stayed about the same. The big change has been Asia and Oceania which has increased its share every year, and this is probably due to the shift in manufacturing to countries like China and India from the West.

	Remember! It's AO3 so you need to check for skills, including statistical skills.		
Strengths of the answer			
Ways to improve the answer			
Level		**Mark**	

Examiner feedback

The candidate correctly identifies the broad changes between 1973 and 2013. The percentage share data are used to illustrate what has happened in specific regions (four regions are named and illustrated). There is a hint that the candidate is making links in the last sentence, recognises that regional shares are all linked to globalisation.

However, much of what the candidate writes is illustrative. There's no manipulation of the data. For example, the candidate doesn't:

- calculate any values, such as the actual amount in US$ for North America in 2013 compared to 1973
- identify that the value of global trade has increased by such a large amount that every region has increased the value of its exports at every year.

Without any manipulation of data (i.e. calculating changes), the candidate cannot progress beyond Level 1. The candidate therefore obtains just 2 marks.

4 Now mark this one!

Read through Sample Answer 2 below.

a) Go through the answer using the three colours in section 3, and underline any links back to the question.
b) Look for evidence of both knowledge and understanding.
c) Use the mark scheme on the previous page to decide which level it is in and how many marks it is worth.

 **Question recap**

Analyse the data in Figure 1.
(6 marks)

Sample Answer 2

The changes in world exports have been staggering, whose value increased by over 31 times between 1973 and 2013. The big growth region of the world has been Asia and Oceania whose share has increased from 14.9% to 31.5%. Europe has the world's largest share of exports at each year, though its share has decreased to 36.3%, and it will probably be overtaken by Asia and Oceania before long. However, when taken by value, while percentage share may have fallen for the majority of regions, the value has risen for every region at every year. So North America's 17.3% share in 1973 by value was US$100 billion, rising to a 13.2% share of US$18 301 – i.e. US$2415 billion, a 24-times increase. This reflects the fact that globalisation has increased trade by value for every region of the world – even Africa, from US$28 billion in 1973 to US$604 billion in 2013.

	Remember! It's AO3 so you need to check for skills, including statistical skills.	
Strengths of the answer		
Ways to improve the answer		
Level		**Mark**	

5 Marked sample answers

Sample Answer 2 is marked below. The text has been highlighted to show how well the answer has been structured.

The following have been highlighted:

- points in red
- explanations in orange
- evidence in blue.

Marked sample answer 2

Point – huge changes in value, quantified by manipulating the data

Explanation – extends the point by explaining how Europe may be overtaken by Asia and Oceania

Evidence – illustrates the point using North America

The changes in world exports have been staggering, whose value increased by over 31 times between 1973 and 2013. The big growth region of the world has been Asia and Oceania whose share has increased from 14.9% to 31.5%. Europe has the world's largest share of exports at each year, though its share has decreased to 36.3%, and it will probably be overtaken by Asia and Oceania before long. However, when taken by value, while percentage share may have fallen for the majority of regions, the value has risen for every region at every year. So North America's 17.3% share in 1973 by value was US$100 billion, rising to a 13.2% share of US$18 301 – i.e. US$2415 billion, a 24-times increase. This reflects the fact that globalisation has increased trade by value for every region of the world – even Africa, from US$28 billion in 1973 to US$604 billion in 2013.

Evidence – uses Asia and Oceania as an example of this growth

Point – identifies the increased value of exports for all regions at each date

Explanation – extends the point by illustrating how even Africa has benefited from increased value

✓ Examiner feedback

The descriptor for Level 2 applies to this answer as follows:

- *Clear analysis of the quantitative evidence provided which makes appropriate use of data to support.* The candidate picks apart both the general trends, and individual regions of the world to illustrate trends, and manipulates the data converting percentage shares into actual amounts.
- *Clear connections between different aspects of the data.* The candidate identifies key trends which help to link the impacts of globalisation between regions, e.g. as North America loses percentage share, so Asia and Oceania increase it.

By meeting both AO3 descriptors fully, the answer earns all 6 marks.

In this section you'll learn how to...

- maximise marks on 6-mark questions that use the command word 'Assess'.

What this section is about

This section helps you to understand what is required in answering 6-mark questions that use the command word 'Assess'. These questions:

- use a resource stimulus (e.g. a photo), the purpose of which is to jog your thoughts
- assess AO1 (knowledge and understanding) for 2 marks, and AO2 (application) for 4 marks. So use what you know and apply it to the resource.

Handy Hint!

You could come across 6-mark questions that use a resource as a stimulus in most questions in Paper 1 and Paper 2.

Five steps to success!

Five steps to help you write top quality answers

The following five steps are used in this book to help you get the best marks.

1. **Plan your answer** – decide what to include and how to structure your answer.

2. **Write your answer** – use the answer spaces to complete your answer.

3. **Mark your answer** – use the mark scheme to self- or peer-mark your answer. You can also use this to assess sample answers in step 4 below.

4. **Sample answers** – two sample answers are given to show you how to maximise marks for a question.

5. **Marked sample answer** – this is the same answer that you used for step 4 above, but is marked and annotated, so that you can compare with your own answers.

Question

Using Figure 1 and your own knowledge, assess the impact of farming practices on the water cycle. (6 marks)

Figure 1 *Spray irrigation being used on a grass crop in New South Wales, Australia*

This question assesses AO1 (knowledge and understanding) for 2 marks and AO2 (application) for 4 marks.

1 Plan your answer

 Tip

Before attempting to answer the question, remember to **BUG** it. That means:

✓ **Box** the command word.
✓ **Underline** the following:
  • the theme
  • the focus
  • any evidence required
  • the number of points needed.
✓ **Glance** back over the question – to make sure you include everything in your answer.

Use the **BUG** below to plan your own answer.

> **Remember, quality not quantity**
>
> You will not be marked on the **number** of points you make, but on the **quality** of your answer. That means the quality of the content, the evidence you give, and how you link back to the question.

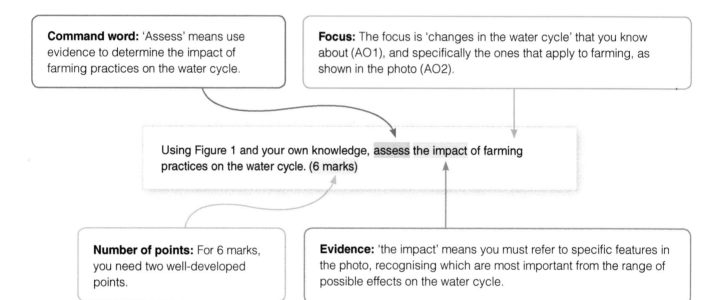

Command word: 'Assess' means use evidence to determine the impact of farming practices on the water cycle.

Focus: The focus is 'changes in the water cycle' that you know about (AO1), and specifically the ones that apply to farming, as shown in the photo (AO2).

Using Figure 1 and your own knowledge, assess the impact of farming practices on the water cycle. (6 marks)

Number of points: For 6 marks, you need two well-developed points.

Evidence: 'the impact' means you must refer to specific features in the photo, recognising which are most important from the range of possible effects on the water cycle.

PEEL your answer

Use PEEL notes to help structure your answer. This will help you explain your ideas to the examiner most clearly. PEEL has four stages:

• **P**oint – Make two well-developed points for 6 marks. Use sentences, not bullet points. These points are the **AO2** parts of the answer where you identify impacts of farming practices which most affect the water cycle.
• **E**vidence – Include details from features shown in the photo. These support your points and explanations.
• **E**xplain – Give a reason for each point you make. Use starter phrases such as: '*This is because ...*', or '*One reason is ...*'. This is the **AO1** part of the answer where you show what you know about farming and the water cycle.
• **L**ink back to the question – Use the question wording in your answer to show the examiner you recognise what it is asking, e.g. '*This process has most influenced the water cycle because ....*'.

Remember the AOs!

Remember how important it is to include elements of both AO1 **and** AO2 in your answer.
• AO1 is your knowledge and understanding – in this case, about the water cycle (e.g. spray irrigation and evaporation rates). But remember, if you do just show what you know (AO1) without referring to the photo (AO2), you can only gain 2 marks.
• AO2 is about applying what you know to the question – in this case:
  a) being able to quote evidence from the photo to show which aspects of farming would affect the water cycle
  b) selecting which aspects of farming might impact most upon the water cycle.

2 Write your answer

Using Figure 1 and your own knowledge, assess the impact of farming practices on the water cycle. (6 marks)

_____

_____

_____

_____

_____

_____

_____

_____

_____

_____

_____

_____

_____

_____

_____

3 Mark your answer

1. To help you to identify if the answer includes well-structured points, first highlight or underline:

- points in red
- explanations in orange
- evidence in blue
- links back to the question.

2. Use the mark scheme on the next page to decide what mark to give.

- Six-mark questions are not marked using individual points, but instead, you should choose a level and a mark based upon the quality of an answer as a whole.
- Remember, a top-level answer must include both AO1 **and** AO2. An answer which includes only AO1, without applying those points to the photo, can only gain a maximum of 3 marks.

Level	Marks	Descriptor	Examples
2	4–6	• Demonstrates clear knowledge and understanding of concepts, processes, interactions and change. (AO1) • Applies knowledge and understanding to the photo, offering clear analysis and evaluation drawn from it. (AO2) • Connections and relationships between different aspects of study are evident with clear relevance. (AO2)	• Spray irrigation is one of the most wasteful ways of applying water during hot dry weather because it evaporates easily. • Irrigation uses ground water, which is often very ancient, for short-term farming gains, thus depleting deep stores in the water cycle and reducing flows between deep stores, unless there are heavy rains later to top it up.
1	1–3	• Demonstrates basic knowledge and understanding of concepts, processes, interactions, change. (AO1) • Applies limited knowledge and understanding to the photo, offering basic analysis and evaluation drawn from it. (AO2) • Connections and relationships between different aspects of study are basic with limited relevance. (AO2)	• Irrigating crops is used by farmers to water crops, but it can be wasteful. • The water used in the photo is a flow out of the water cycle system, so it is lost and you can't use it again. This is called having a water deficit.
0	0	Nothing worthy of credit.	

Things to watch out for

Study Sample Answer 1 below.

• Check that the candidate knows aspects of the water cycle.
• Check that the candidate is referring to the photograph as demanded by the question.

Clues:
a) Look through the answer for AO1, i.e. what the candidate knows, and check the candidate is answering the question. Underline any AO1 points.
b) Then look for AO2, i.e. how well the candidate applies what they know to what is in the photo. Highlight any AO2 points.

Sample Answer 1

The photo shows spray irrigation, which is being used to cultivate a crop. This is used in climates that are hot and dry or during dry summer spells when soils are too dry. The crop is probably grass which is used as winter feed for cattle, so the water is essential to maintain plant growth. Water abstraction is used because surface streams might have dried up. This water is most likely coming from a borehole and, since the photo is in Australia, then it might be artesian water from deep underground. Farmers rely on this water during droughts. Although abstraction would lower the water table deep underground, it would rise again when rains come.

	Remember! Check for both AO1 (knowledge and understanding) and AO2 (application).		
Strengths of the answer			
Ways to improve the answer			
Level		**Mark**	

Examiner feedback

The candidate identifies the role of irrigation. However, the answer is a simple recall of what the candidate knows about irrigation – there's nothing that connects what is happening in the photo to the water cycle as a whole.

- The question asks for an assessment of the impact of farming practices and water abstraction on the water cycle. What the candidate gives is limited knowledge and understanding of why the farmer is irrigating this land.
- There is nothing that applies this limited understanding to the photo or shows the impact on the water cycle until the last sentence – and, even then, it does not assess how big or small this impact may be.

Without this, the candidate cannot progress to Level 2, either for AO1, or AO2. The candidate therefore obtains a 'best fit' of just 2 marks.

4 Now mark this one!

Read through Sample Answer 2 below.

a) Go through the answer using the three colours in section 3, and underline any links back to the question.
b) Remember, any points identifying features about the photo and knowledge about the water cycle or irrigation are AO1 (knowledge and understanding), and applying these to the water cycle and a systems approach are AO2 (application).
c) Use the mark scheme to decide how many marks it is worth.

 Question recap

Using Figure 1 and your own knowledge, assess the impact of farming practices on the water cycle. (6 marks)

Sample Answer 2

The photo shows spray irrigation, commonly used either during drought, or as a top-up to water crops. It comes from deep underground storage in the water cycle, and prolonged use can reduce storage levels and therefore groundwater flow into streams and rivers. If drought is prolonged then spray irrigation is a benefit for farmers in the short term, but may be creating long-term problems by reducing storage.

The photo is from Australia, where drought is more common than the UK. Much spray irrigation would be evaporated as an output from the water cycle, and therefore lost into the atmosphere without benefiting the plants that it is being used for. Compared to other methods of irrigation, it is therefore wasteful. Much of the moisture would only be retained briefly in the soil before being drawn up by transpiration via roots, and being lost as an output.

	Remember! Check for both AO1 (knowledge and understanding) and AO2 (application).		
Strengths of the answer			
Ways to improve the answer			
Level		**Mark**	

5 Marked sample answer

Sample Answer 2 is marked below. The text has been highlighted to show how well the answer has been structured.

The following have been highlighted:

- points in red
- explanations in orange, particularly of the impact of the water cycle
- evidence in blue, especially of knowledge of aspects of the water cycle
- underlined points are where the candidate links to the question to assess the impact of farming practices on the water cycle.

Marked sample answer 2

Evidence from photo – identifies spray irrigation and a problem in using it

Explanation – extends the point by expanding on the impacts of prolonged spray irrigation

Point – identifies wastefulness of spray irrigation

> The photo shows spray irrigation, commonly used either during drought, or as a top-up to water crops. It comes from deep underground storage in the water cycle, and prolonged use can reduce storage levels and therefore groundwater flow into streams and rivers. If drought is prolonged then spray irrigation is a benefit for farmers in the short term, but may be creating long-term problems by reducing storage.
>
> The photo is from Australia, where drought is more common than the UK. Much spray irrigation would be evaporated as an output from the water cycle, and therefore lost into the atmosphere without benefiting the plants that it is being used for. Compared to other methods of irrigation, it is therefore wasteful. Much of the moisture would only be retained briefly in the soil before being drawn up by transpiration via roots, and being lost as an output.

Point – assesses the impact of this on the water cycle

Evidence from photo – knowledge of Australia and evaporation resulting from spray irrigation

Explanation – extends the point by explaining losses from transpiration

Links – all the underlined points in the answer are where the candidate addresses the focus of question to assess 'the impact on the water cycle'.

  Examiner feedback

The descriptor for Level 2 applies fully to this answer as follows:

- For AO1:
  Demonstrates clear knowledge and understanding of concepts, processes, interactions and change. The candidate shows excellent knowledge about farming practices processes using spray irrigation and terminology of the water cycle.
- For AO2:
  Applies knowledge and understanding to the photo, offering clear analysis and evaluation drawn from it. Connections and relationships between different aspects of study are evident with clear relevance. There is clear reference to the photo and to features in the photo which link to the water cycle. The candidate assesses the impact on the water cycle fully.

By meeting both assessment objectives descriptors fully, the answer earns all 6 marks. It's a top-quality answer.

In this section you'll learn how to...

- maximise marks on 9-mark questions on Hazards, using the command word 'Assess'.

Tackling 9-mark questions about Hazards

This topic is assessed on Paper 1, Section C, Question 5.

- Before beginning this section, read section 1.6, which will tell you about the skills you need to answer 9-mark questions, and the mark scheme.

Try this 9-mark question on Hazards

Question

Assess the reasons why earthquakes of similar magnitude can have different impacts. (9 marks)

This question assesses AO1 (knowledge and understanding) for 4 marks and AO2 (application) for 5 marks.

1 Plan your answer

Before attempting to answer the question, remember to **BUG** it. That means:

✓ **Box** the command word.
✓ **Underline** the following:
  - the theme
  - the focus
  - any evidence required
  - the number of points needed.
✓ **Glance** back over the question – to make sure you include everything in your answer.

Use the **BUG** on the next page to plan your own answer.

Five steps to success!

Five steps to help you write top quality answers

The following five steps are used in this book to help you get the best marks.

1. **Plan your answer** – decide what to include and how to structure your answer.

2. **Write your answer** – use the answer spaces to write your answer.

3. **Mark your answer** – use the mark scheme (section 1.6) to self- or peer-mark your answer. You can also use this to assess sample answers in step 4 below.

4. **Sample answers** – two sample answers are given to show you how to maximise marks for a question.

5. **Marked sample answer** – this is the same answer that you used for step 4 above, but is marked and annotated, so that you can compare with your own answers.

Command word: 'Assess' means use evidence to determine why earthquakes of similar magnitude have different impacts, saying why some reasons are most important in explaining these impacts.

Focus: The focus is earthquakes from *Hazards*, so you need to know something about earthquake impacts, and why they vary in different locations.

Assess the reasons why earthquakes of similar magnitude can have different impacts. (9 marks)

Number of points: 9 marks. You need three well-developed points, each helping to develop the argument (AO2), and supported by evidence (AO1).

Evidence: 'reasons why earthquakes of similar magnitude can have different impacts' means you must use evidence from different earthquakes to explain why this is the case

PEEL your answer

Use PEEL notes to structure your answer. This will help to explain your ideas to the examiner clearly. PEEL has four stages:

* **P**oint – Make three well-developed points for 9 marks. These are the **AO2** parts of the answer where you argue how important particular factors are, and how they explain why impacts of earthquakes of similar magnitude vary.
* **E**vidence – Include details about specific earthquakes to illustrate your argument. It's important to quote examples, such as the magnitude of the 2004 Indonesian earthquake. This is **AO1** – using knowledge to support your argument.
* **E**xplain – Give a reason for each point. Use starter phrases such as: '*The lack of warning system at .... helps to explain why this earthquake had such great impacts because ...*'. This extends your **AO2** ability because you'll show why earthquakes of similar magnitude have different impacts.
* **L**ink back to the question – Use the wording of the question in your answer to show you recognise what the question is asking, e.g. '*This example shows why earthquakes of similar magnitude can have different impacts because ...*'.

 Tip

Remember, quality not quantity

You will not be marked on the **number** of points you make, but on the **quality** of your answer. That means the quality of the argument (AO2), the evidence you give (AO1), and how you link back to the question.

Remember the AOs!

It's important to know the AOs for this question.
* **AO1** (4 marks) is about showing knowledge and understanding of specific earthquakes, and of their impacts. Use brief examples – not long case studies.
* **AO2** (5 marks) is about using evidence to decide the factors which explain why earthquakes of similar magnitude have different impacts.

② Write your answer

Assess the reasons why earthquakes of similar magnitude can have different impacts. (9 marks)

_____

_____

_____

_____

_____

_____

_____

_____

_____

_____

_____

_____

_____

_____

_____

_____

_____

_____

_____

_____

_____

You can continue on file paper if you need more space

③ Mark your answer

1. To help you to identify well-structured points, highlight or underline:

- points in red
- explanations in orange
- evidence in blue
- links back to the question by underlining.

2. Use the mark scheme in section 1.6. Remember that:

- 9-mark questions are marked by choosing a level and a mark within it, based upon the answer as a whole.
- High-scoring answers must include both AO1 **and** AO2. One which includes only AO1, without explaining the answer for AO2, can only gain a maximum of 4 marks.

Things to watch out for

Study Sample Answer 1 below.

- The answer has strengths, so identify them.
- Suggest what prevents it from getting a higher mark.

Clues:
Read how the candidate demonstrates:
a) **AO1** – what does the candidate know about particular earthquakes? their impacts?
b) **AO2** – how well does the candidate explain why the earthquakes had different impacts?

Sample Answer 1

There are several reasons why earthquakes of similar magnitude can have very different impacts. The two earthquakes that affected Indonesia on Boxing Day 2004 (known for its major tsunami) and Tōhoku in Japan in 2011 were similar in magnitude. Indonesia was slightly bigger (estimated between 9 and 9.3 on the Richter scale) than Japan (9) but they are two of the greatest earthquakes in the last century. However they had different impacts.

The earthquake in Indonesia caused the largest tsunami in recent history. The earth movement was 15 metres along the subduction zone between the Indian and Burma plates which started off the tsunami. This killed over 300 000 people in countries like Malaysia, Thailand, Sri Lanka. The closer you were to the earthquake, the worse the economic impacts, for example 44% of the population of Aceh province in Indonesia lost their livelihoods. Over 31 000 people were killed in Sri Lanka compared to 235 000 in Indonesia.

The earthquake in Tōhoku in Japan was like the one in Indonesia because the biggest impacts were from the tsunami. Over 18 000 people died, which is less than the Indonesian earthquake. Half a million people were made homeless. Minami–Sanriku was the worst affected place where 50% of the deaths occurred. But some places were hardly affected. The Fukushima nuclear power station exploded when the tsunami reached it, causing radiation leaks. The population had to be evacuated and engineers feared the reactors would melt down.

	Remember! Check for both AO1 (knowledge and understanding) and AO2 (application).	
Strengths of the answer		
Ways to improve the answer		
Level		**Mark**	

Examiner feedback

The answer has strengths.

- Earthquakes are accurately described, and the candidate shows knowledge and understanding of each. (AO1)
- However, the candidate does not fully 'explain' why impacts were so different. (AO2)
- To earn a higher mark, the candidate needs to develop explanations to say why these earthquakes had such different impacts. (AO2)

The answer therefore fits criteria for different levels!

For AO1, the candidate fits the criterion for Level 3:

- There is evidence of *detailed knowledge*, of *clear understanding of concepts*, and of *physical and human processes*. Evidence of these can be found throughout the answer.

However, AO2 is only Level 1–2:

- *Knowledge and understanding* is rarely applied to the question. (Level 1)
- *Links between different parts of the answer* are not expanded upon. (Level 1).
- There is evidence of *general analysis, supported with some evidence*. (Level 2)
- There is little to say *why the earthquakes varied*. (Level 1)

The answer therefore gets a 'best fit' mark of Level 2, with 4 marks.

4 Now mark this one!

Read through Sample Answer 2 below.

a) Go through the answer using the three colours in section 3, and underline any links back to the question.
b) Look for evidence of both knowledge and understanding.
c) Use the mark scheme in section 1.6 to decide on a level and mark.

 **Question recap**

Assess the reasons why earthquakes of similar magnitude can have different impacts.
(9 marks)

Sample Answer 2

There are several reasons why earthquakes of similar magnitude can have very different impacts. For example, the earthquake that affected Indonesia on Boxing Day 2004 (and which caused the Asian tsunami) contrasted with that in Tōhoku, Japan in 2011. The two were similar magnitude – each about 9 on the Richter scale, but each had different impacts.

The Asian tsunami was caused by a vertical earth movement of 15 metres along the subduction zone between the Indian and Burma plates. By far the most important reason for the devastating impact of this tsunami was the lack of any early warning system. Within 15 minutes of the earthquake, the tsunami wave had struck Sumatra, Indonesia, killing over 230 000 people, with no advance warning. The closer to the focus of the earthquake, the worse the impacts; the city of Banda Aceh in Indonesia was virtually obliterated; no warning system could have prevented that. 44% of the population of Aceh province lost their livelihoods. Physical distance was an important factor in the varied impacts between countries; although 31 000 people were killed in Sri Lanka, the toll was much less than in Indonesia because two hours warning had been given.

The earthquake in Tōhoku in Japan in 2011 was similar because the worst impacts were nearest the epicentre. Like Indonesia, the biggest impacts were from the tsunami. But Japan has warning systems using alarms, civil defence, and the media, as well as a sophisticated electronic system created by detection of tsunami waves at sea. Over 18 000 people died, which is much less than the Indonesian earthquake because they had warnings, and most of those who died had only 30 minutes warning or less. Half a million people were made homeless, and 50% of all deaths occurred in Minami-Sanriku near the epicentre.

But the worst impact had nothing to do with warning or physical distance – it occurred when reactors at the Fukushima nuclear power station exploded and caused radiation leaks, and evacuation of the population. The most important reason why this occurred was that the tsunami barriers along the coast were too small.

	Remember! Check for both AO1 (knowledge and understanding) and AO2 (application).		
Strengths of the answer			
Ways to improve the answer			
Level		**Mark**	

5 Marked sample answers

Sample Answer 2 is marked below. The text has been highlighted to show how well each answer has structured points.

The following have been highlighted:

- points in red
- explanations in orange
- evidence in blue
- underlined points are where the candidate links to the question to assess reasons why the earthquakes varied.

Marked sample answer 2

Link – introduces the answer using the question wording, and locates the examples

There are several reasons why earthquakes of similar magnitude can have very different impacts. For example, the earthquake that affected Indonesia on Boxing Day 2004 (and which caused the Asian tsunami) contrasted with that in Tōhoku, Japan in 2011. The two were similar magnitude – each about 9 on the Richter scale, but each had different impacts.

The Asian tsunami was caused by a vertical earth movement of 15 metres along the subduction zone between the Indian and Burma plates.

Point – names the cause of the Asian tsunami

Explanation – identifies the most important factor in explaining impacts of the tsunami

Evidence – provides evidence about the impacts about the tsunami

Point – explains where impacts were greatest, i.e. nearest the epicentre

Explanation – extends the point by showing the importance of warnings to the death toll

Point – emphasises the nuclear power station explosions as a key factor in explain impacts of the tsunami

By far the most important reason for the devastating impact of this tsunami was the lack of any early warning system. Within 15 minutes of the earthquake, the tsunami wave had struck Sumatra, Indonesia, killing over 230 000 people, with no advance warning. The closer to the focus of the earthquake, the worse the impacts; the city of Banda Aceh in Indonesia was virtually obliterated; no warning system could have prevented that. 44% of the population of Aceh province lost their livelihoods. Physical distance was an important factor in the varied impacts between countries; although 31 000 people were killed in Sri Lanka, the toll was much less than in Indonesia because two hours warning had been given.

The earthquake in Tōhoku in Japan in 2011 was similar because the worst impacts were nearest the epicentre. Like Indonesia, the biggest impacts were from the tsunami. But Japan has warning systems using alarms, civil defence, and the media, as well as a sophisticated electronic system created by detection of tsunami waves at sea. Over 18 000 people died, which is much less than the Indonesian earthquake because they had warnings, and most of those who died had only 30 minutes warning or less. Half a million people were made homeless, and 50% of all deaths occurred in Minami-Sanriku near the epicentre.

But the worst impact had nothing to do with warning or physical distance – it occurred when reactors at the Fukushima nuclear power station exploded and caused radiation leaks, and evacuation of the population. The most important reason why this occurred was that the tsunami barriers along the coast were too small.

Link – the explanation links back to the question by identifying a key factor

Link – links back to the question by identifying a second factor

Explanation – extends the point by showing how most people were warned

Evidence – provides evidence of impacts close to epicentre

Link – links back to the question by stating that these factors are less important than nuclear power station explosions

Link – links back to the question by identifying the key reason why explosions occurred

Explanation – explains reason for the explosions

✓ **Examiner feedback**

This is a very strong answer. The descriptor for Level 3 applies to this answer as follows:

For AO1:

- *There is evidence of detailed knowledge, of clear understanding of concepts, and of physical and human processes throughout the answer.*

For AO2:

- *Applies knowledge and understanding of geographical information/ideas logically, making relevant connections/relationships.*
- *Applies knowledge and understanding of geographical information/ideas to produce a full and coherent interpretation that is relevant and supported by evidence.*
- *Applies knowledge and understanding of geographical information/ideas to make supported judgements about the significance of factors throughout the response, leading to a balanced and coherent argument.*

There are five underlined links back to the question – all of which show that the candidate can 'assess'. By meeting the Level 3 descriptors fully, the answer earns the full 9 marks.

In this section you'll learn how to...

- maximise marks on 9-mark questions on Ecosystems under stress, using the command term 'To what extent?'.

Tackling 9-mark questions about Ecosystems under stress

This topic is assessed on Paper 1, Section C, Question 6.

- Before beginning this section, read section 1.6, which will tell you about the skills you need to answer 9-mark questions, and the mark scheme.

Try this 9-mark question on Ecosystems under stress

> This question assesses AO1 (knowledge and understanding) for 4 marks and AO2 (application) for 5 marks.

Question

To what extent do human factors explain the ecological change experienced by one region that you have studied? (9 marks)

1 Plan your answer

Before attempting to answer the question, remember to **BUG** it. That means:

✓ **Box** the command word.
✓ **Underline** the following:
  - the theme
  - the focus
  - any evidence required
  - the number of points needed.
✓ **Glance** back over the question – to make sure you include everything in your answer.

Use the **BUG** on the next page to plan your own answer.

Five steps to success!

Five steps to help you write top quality answers

The following five steps are used in this book to help you get the best marks.

1. **Plan your answer** – decide what to include and how to structure your answer.

2. **Write your answer** – use the answer spaces to write your answer.

3. **Mark your answer** – use the mark scheme (section 1.6) to self- or peer-mark your answer. You can also use this to assess sample answers in step 4 below.

4. **Sample answers** – two sample answers are given to show you how to maximise marks for a question.

5. **Marked sample answer** – this is the same answer that you used for step 4 above, but is marked and annotated, so that you can compare with your own answers.

Command term: 'To what extent' means to use evidence to determine and weigh up the ways in which human factors have caused ecological change within one region, and reach a judgement.

Evidence: The extent to which 'human factors explain the ecological change' means you must use evidence from one region to explain how far human factors are responsible for the changes.

To what extent do human factors explain the ecological change experienced by one region that you have studied? (9 marks)

Focus: The focus is a regional study, so you need to know which region you can write about which is changing ecologically, and to what extent human factors explain these changes.

Number of points: 9 marks. You need three well-developed points, each helping to develop the argument (AO2), and supported by evidence (AO1).

PEEL your answer

Use PEEL notes to structure your answer. This will help to explain your ideas to the examiner clearly. PEEL has four stages:

- **P**oint – Make three well-developed points for 9 marks. These are the **AO2** parts of the answer where you argue that ecological changes are occurring, and to what extent human factors explain these.
- **E**vidence – Include details about specific ecological changes to illustrate your argument. It's important to quote examples, such as overgrazing in the Sahel. This is **AO1** – using knowledge to support your argument.
- **E**xplain – Give a reason for each point. Use starter phrases such as: '*The use of fire to clear savanna grassland explains the changes taking place because ...*'. This extends your **AO2** ability because you'll show the extent to which human factors explain ecological change.
- **L**ink back to the question – Use the wording of the question to show you recognise what the question is asking, e.g. '*This example shows how important human factors can cause ecological change ...*'.

 Tip

Remember, quality not quantity

You will not be marked on the **number** of points you make, but on the **quality** of your answer. That means the quality of the argument (AO2), the evidence you give (AO1), and how you link back to the question.

Remember the AOs!

It's important to know the AOs for this question.
- **AO1** (4 marks) is about showing knowledge and understanding of specific ecological changes, and of their impacts. Use brief examples – not long case studies.
- **AO2** (5 marks) is about using evidence to decide the extent to which human factors explain these changes.

2   Write your answer

To what extent do human factors explain the ecological change experienced by one region that you have studied? (9 marks)

_____

_____

_____

_____

_____

_____

_____

_____

_____

_____

_____

_____

_____

_____

_____

_____

_____

_____

_____

_____

_____

You can continue on file paper if you need more space

3   Mark your answer

1. To help you to identify well-structured points, highlight or underline:

- points in red
- explanations in orange
- evidence in blue
- links back to the question by underlining.

2. Use the mark scheme in section 1.6. Remember that:

- 9-mark questions are marked by choosing a level and a mark within it, based upon the answer as a whole.
- High-scoring answers must include **both** AOs. One which includes only AO1, without explaining the answer for AO2, can only gain a maximum of 4 marks.

Things to watch out for

Study Sample Answer 1 below.

- The answer has strengths, so identify them.
- Suggest what prevents it from getting a higher mark.

Clues:
Read how the candidate demonstrates:
a) **AO1** – what does the candidate know about ecological changes and human factors involved in these changes?
b) **AO2** – how well does the candidate explain the extent to which human factors explain these changes?

Sample Answer 1

Savanna grasslands are changing from farming, when cattle overgraze grassland areas. Nomadic peoples used to look after the land because their cattle would not stay long enough to damage the grasses, but population pressure has changed this and more cattle are being kept permanently in one place. This causes overgrazing, especially when there is a long drought. Overgrazing often destroys grass roots, so when the wind blows, soil is eroded and blown away. An example of a region affected like this is the Sahel.

Another problem is reduced biodiversity where cattle have overgrazed. During droughts, cattle will eat tree leaves or bark to survive when grass dies, so it means that trees are stripped bare and die. Only a few species survive if they have long thorns that prevent cattle from eating branches and leaves. So these are the species which seed and spread, while others die back.

Tourism also causes changes to the savanna. People want to see the savanna wildlife such as giraffe or elephant, so they focus on those areas. Before long, roads and hotels have developed and litter has become a problem. The use of safari 4x4 vehicles can cause severe soil erosion, especially in the rainy season.

	Remember! Check for both AO1 (knowledge and understanding) and AO2 (application).	
Strengths of the answer		
Ways to improve the answer		
Level		**Mark**	

Examiner feedback

The answer has strengths.

- The candidate shows knowledge and understanding of the ecology of the savanna, and changes that it faces. (AO1)
- However, the candidate does not explain 'to what extent' human factors cause these changes (AO2). It is clear that human factors ARE causing change, but the candidate doesn't say so.
- To earn a higher mark, the candidate needs to explain to what extent human factors causes ecological changes. (AO2)

The answer fits criteria for different levels. For AO1, the candidate fits the criterion for Level 3:

- There is evidence of *detailed knowledge,* of *clear understanding of concepts,* and of *physical and human processes.* Evidence of these can be found throughout the answer.

However, AO2 is only Level 1–2:

- *Knowledge and understanding* is rarely applied to the question. (Level 1)
- *Links between different parts of the answer* are not expanded upon. (Level 1)
- There is evidence of *general analysis, supported with some evidence.* (Level 2)
- There is little to say *to what extent humans causes the ecological changes.* (Level 1)

The answer therefore gets a 'best fit' mark of Level 2, with 4 marks.

4 Now mark this one!

Read through Sample Answer 2 below.

- a) Go through the answer using the three colours in section 3, and underline any links back to the question.
- b) Look for evidence of both knowledge and understanding.
- c) Use the mark scheme in section 1.6 to decide on a level and mark.

 Question recap

To what extent do human factors explain the ecological change experienced by one region that you have studied? (9 marks)

Sample Answer 2

Savanna regions are changing, caused mostly by human factors. Drought, a natural cause, has affected the African savanna for over 50 years, which has killed some species of grasses and smaller, shallow-rooted trees, but human factors have caused greater change.

The biggest factor is population pressure, leading to overgrazing. Nomadic peoples in the Sahel used to move with their cattle, which would not stay in one place for long enough to damage the ecosystem. Now population pressure has led to more cattle being kept permanently in one place. More people keep cattle as a status and wealth symbol in countries such as Botswana, where overgrazing has destroyed grass roots, causing soil erosion by wind.

A further problem caused by overgrazing is reduced biodiversity. Increased cattle numbers have added to problems caused by drought, because cattle eat tree leaves or bark to survive as grass dies. This means that trees are stripped bare and die, so only a few species can survive if they have long thorns that prevent cattle from eating the branches and leaves. These are the species which survive and spread in the Sahel. Drought started the problem, but human factors made it worse.

Tourism is a final human factor. People visit to see wildlife such as giraffe or elephant. This leads to more roads and hotels in wildlife hotspots, and litter becomes a problem in Kenya and Tanzania. The use of safari 4x4 vehicles has been blamed for severe soil erosion, especially in the rainy season.

	Remember! Check for both AO1 (knowledge and understanding) and AO2 (application).		
Strengths of the answer			
Ways to improve the answer			
Level		**Mark**	

5 Marked sample answers

Sample Answer 2 is marked below. The text has been highlighted to show how well each answer has structured points.

The following have been highlighted:

- points in red
- explanations in orange
- evidence in blue
- underlined points are where the candidate links to the question to show the influence of human factors.

Marked sample answer 2

Link – uses the question wording to begin the argument

Point – introduces the balance between physical factors (drought) and human

Point – identifies population pressure as the greatest problem

Explanation – explains how the changes have taken place over time

Evidence – uses the example of Botswana to support the point

Savanna regions are changing, caused mostly by human factors. Drought, a natural cause, has affected the African savanna for over 50 years, which has killed some species of grasses and smaller, shallow-rooted trees, but human factors have caused greater change.

The biggest factor is population pressure, leading to overgrazing. Nomadic peoples in the Sahel used to move with their cattle, which would not stay in one place for long enough to damage the ecosystem. Now population pressure has led to more cattle being kept permanently in one place. More people keep cattle as a status and wealth symbol in countries such as Botswana, where overgrazing has destroyed grass roots, causing soil erosion by wind.

Point – identifies reduced biodiversity as a consequence of overgrazing

Evidence – provides evidence from the Sahel

Link – uses wording of the question to identify tourism as a problem

Explanation – explains how overgrazing has affected biodiversity

Link – links back to the question to show human factors made things worse

Point – identifies wildlife as the attraction

A further problem caused by overgrazing is reduced biodiversity. Increased cattle numbers have added to problems caused by drought, because cattle eat tree leaves or bark to survive as grass dies. This means that trees are stripped bare and die, so only a few species can survive if they have long thorns that prevent cattle from eating the branches and leaves. These are the species which survive and spread in the Sahel. Drought started the problem, but human factors made it worse.

Tourism is a final human factor. People visit to see wildlife such as giraffe or elephant. This leads to more roads and hotels in wildlife hotspots, and litter becomes a problem in Kenya and Tanzania. The use of safari 4x4 vehicles has been blamed for severe soil erosion, especially in the rainy season.

Evidence – provides evidence from Kenya and Tanzania

Explanation – extends the point using the example of 4x4 vehicles

 Examiner feedback

This is a very strong answer. The descriptor for Level 3 applies to this answer as follows:

For AO1:

- *There is evidence of detailed knowledge, of clear understanding of concepts, and of physical and human processes throughout the answer.*

For AO2:

- *Applies knowledge and understanding of geographical information/ideas logically, making relevant connections/relationships.*
- *Applies knowledge and understanding of geographical information/ideas to produce a full and coherent interpretation that is relevant and supported by evidence.*
- *Applies knowledge and understanding of geographical information/ideas to make supported judgements about the significance of factors throughout the response, leading to a balanced and coherent argument.*

The underlined links back to the question demonstrate that the candidate can show 'to what extent'. By meeting the Level 3 descriptors fully, the answer earns all 9 marks.

In this section you'll learn how to...

- maximise marks on 9-mark questions on Contemporary urban environments, using the command word 'Evaluate'.

Tackling 9-mark questions about Contemporary urban environments

This topic is assessed on Paper 2, Section C, Question 3.

- Before beginning this section, read section 1.6, which will tell you about the skills you need to answer 9-mark questions, and the mark scheme.

Try this 9-mark question on Contemporary urban environments

> This question assesses AO1 (knowledge and understanding) for 4 marks and AO2 (application) for 5 marks.

Question

Evaluate contrasting approaches to waste disposal in one urban area that you have studied. (9 marks)

1 Plan your answer

Before attempting to answer the question, remember to **BUG** it. That means:

✓ **Box** the command word.
✓ **Underline** the following:
  - the theme
  - the focus
  - any evidence required
  - the number of points needed.
✓ **Glance** back over the question – to make sure you include everything in your answer.

Use the **BUG** on the next page to plan your own answer.

Five steps to success!

Five steps to help you write top quality answers

The following five steps are used in this book to help you get the best marks.

1. **Plan your answer** – decide what to include and how to structure your answer.

2. **Write your answer** – use the answer spaces to write your answer.

3. **Mark your answer** – use the mark scheme (section 1.6) to self- or peer-mark your answer. You can also use this to assess sample answers in step 4 below.

4. **Sample answers** – two sample answers are given to show you how to maximise marks for a question.

5. **Marked sample answer** – this is the same answer that you used for step 4 above, but is marked and annotated, so that you can compare with your own answers.

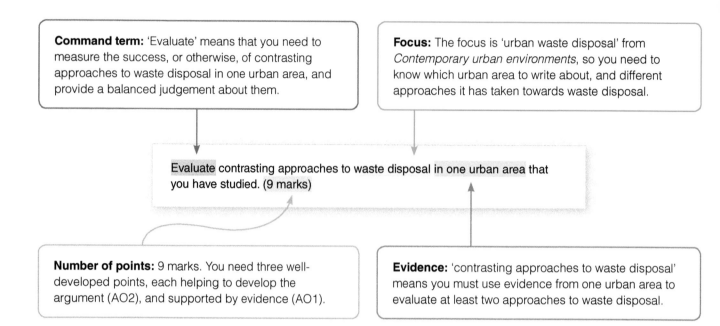

> **Command term:** 'Evaluate' means that you need to measure the success, or otherwise, of contrasting approaches to waste disposal in one urban area, and provide a balanced judgement about them.

> **Focus:** The focus is 'urban waste disposal' from *Contemporary urban environments*, so you need to know which urban area to write about, and different approaches it has taken towards waste disposal.

> Evaluate contrasting approaches to waste disposal in one urban area that you have studied. (9 marks)

> **Number of points:** 9 marks. You need three well-developed points, each helping to develop the argument (AO2), and supported by evidence (AO1).

> **Evidence:** 'contrasting approaches to waste disposal' means you must use evidence from one urban area to evaluate at least two approaches to waste disposal.

PEEL your answer

Use PEEL notes to structure your answer. This will help to explain your ideas to the examiner clearly. PEEL has four stages:

- **P**oint – Make three well-developed points for 9 marks. These are the **AO2** parts of the answer where you argue how effective or successful two or more different waste disposal strategies have been.
- **E**vidence – Include details about specific waste disposal methods to illustrate your argument. It's important to quote examples, such as expanding recycling. This is **AO1** – using knowledge to support your argument.
- **E**xplain – Give a reason for each point. Use starter phrases such as: '*The strategy of expanding landfill sites was unlikely to work because ...*'. This extends your **AO2** ability because you'll explain how effective waste disposal strategies have been.
- **L**ink back to the question – Use the wording of the question in your answer to show you recognise what the question is asking, e.g. '*Increased recycling in London has been a really effective approach to waste disposal because ...*'.

 Tip

Remember, quality not quantity

You will not be marked on the **number** of points you make, but on the **quality** of your answer. That means the quality of the argument (AO2), the evidence you give (AO1), and how you link back to the question.

Remember the AOs!

It's important to know the AOs for this question.
- **AO1** (4 marks) is about showing knowledge and understanding of specific methods of waste disposal. Use brief examples – not long case studies.
- **AO2** (5 marks) is about using evidence to evaluate or judge the success of these methods.

② Write your answer

Evaluate contrasting approaches to waste disposal in one urban area that you have studied. (9 marks)

_____

_____

_____

_____

_____

_____

_____

_____

_____

_____

_____

_____

_____

_____

_____

_____

_____

_____

_____

_____

_____

_____

You can continue on file paper if you need more space

③ Mark your answer

1. To help you to identify well-structured points, highlight or underline:

- points in red
- explanations in orange
- evidence in blue
- links back to the question by underlining.

2. Use the mark scheme in section 1.6. Remember that:

- 9-mark questions are marked by choosing a level and a mark within it, based upon the answer as a whole.
- High-scoring answers must include **both** AOs. One which includes only AO1, without explaining the answer for AO2, can only gain a maximum of 4 marks.

Things to watch out for

Study Sample Answer 1 below.

- The answer has strengths, so identify them.
- Suggest what prevents it from getting a higher mark.

Clues:
Read how the candidate demonstrates:
a) **AO1** – what does the candidate know about contrasting waste policies in a named city?
b) **AO2** – how well does the candidate evaluate the extent to which these have proved successful?

Sample Answer 1

Bristol is a good example of a UK city which has different approaches to waste disposal. It will probably have a population of half a million by 2030, so it needs to do something about waste. It has different policies to deal with waste. It has reduced the amount of waste produced by each household by 15%. It reduced landfill from 400kg per person per year in 2004–05 to nothing by 2012–13. Now the Avonmouth waste treatment plant processes 200 000 tonnes of municipal solid waste each year before the waste goes to an incineration plant which produces electricity for 25 000 homes.

Another policy has been to increase the amount of waste which is recycled from about 50 kg per person per year in 2004–05 to nearly 200 kg per person per year in 2017–18. This is 50% of all waste in the city. This also means there is less going to landfill. The city council collects different materials for recycling at the kerbside, so people don't have to walk or take their recycling. Schools also have recycling policies and candidates are taught about recycling. The recycling and waste collection companies also have targets to meet in what they collect from people's houses.

	Remember! Check for both AO1 (knowledge and understanding) and AO2 (application).		
Strengths of the answer			
Ways to improve the answer			
Level		**Mark**	

Examiner feedback

The answer has strengths.

- The candidate shows knowledge and understanding of different waste disposal policies in Bristol. (AO1)
- However, the candidate does not 'evaluate' these policies (AO2). It is clear that waste disposal policies ARE having results – but the candidate doesn't actually say so, even though the evidence is there.
- To earn a higher mark, the candidate needs to explain how far the policies have been successful. (AO2)

The answer fits criteria for different levels. For AO1, the candidate fits the criterion for Level 3:

- There is evidence of *detailed knowledge*, of *clear understanding of concepts*, and of *physical and human processes*. Evidence of these can be found throughout the answer.

However, AO2 is only Level 1–2:

- *Knowledge and understanding* is rarely applied to the question. (Level 1)
- *Links between different parts of the answer* are not expanded upon. (Level 1)
- There is evidence of *general analysis, supported with some evidence*. (Level 2)
- There is little *evaluation of the success of waste disposal policies*. (Level 1)

The answer therefore gets a 'best fit' mark of mid-Level 2, with 5 marks. This answer could easily be much better with only a few more key words about 'success'.

4 Now mark this one!

Read through Sample Answer 2 below.

a) Go through the answer using the three colours in section 3, and underline any links back to the question.
b) Look for evidence of both knowledge and understanding.
c) Use the mark scheme in section 1.6 to decide on a level and mark.

 Question recap

Evaluate contrasting approaches to waste disposal in one urban area that you have studied.
(9 marks)

Sample Answer 2

Bristol has developed different approaches to waste disposal. Its first success has been to reduce the amount of waste produced by each household by 15% since 2004–05. It has been even more successful in reducing landfill from 400 kg per person per year in 2004–05 to nothing by 2012–13. Its success comes from developing a waste treatment plant at Avonmouth which processes 200 000 tonnes of municipal solid waste each year. This separates recycled waste from what can be burned, so nothing goes to landfill. The remaining waste goes to an incineration plant producing electricity for 25 000 homes.

The second successful policy has been to increase recycling volumes from 50 kg per person per year in 2004–05 to 200 kg per person per year in 2017–18. Half of all waste is now recycled in Bristol, again reducing landfill. The city council collects different materials for recycling at the kerbside to make it easy to recycle. This is reinforced in schools which have recycling policies in the curriculum.

The final policy encourages recycling and waste companies to increase what they collect from people's houses by using targets. Bristol City Council gives specialised targets for particular materials (e.g. composted waste, plastics, or bottles) so this increases their motivation to collect all the recycling they can. This puts responsibility on to companies when they get the contract to collect waste and has led to increases in all types of recycling.

	Remember! Check for both AO1 (knowledge and understanding) and AO2 (application).		
Strengths of the answer			
Ways to improve the answer			
Level		**Mark**	

5 Marked sample answers

Sample Answer 2 is marked below. The text has been highlighted to show how well each answer has structured points.

The following have been highlighted:

- points in red
- explanations in orange
- evidence in blue
- underlined points are where the candidate links to the question to show the influence of human factors.

Marked sample answer 2

Point – the first point is how Bristol has reduced the amount of waste produced

Explanation – extends the point by explaining how waste is now incinerated

Evidence – uses evidence from Bristol to support the point

Bristol has developed different approaches to waste disposal. Its first success has been to reduce the amount of waste produced by each household by 15% since 2004–05. It has been even more successful in reducing landfill from 400 kg per person per year in 2004–05 to nothing by 2012–13. Its success comes from developing a waste treatment plant at Avonmouth which processes 200 000 tonnes of municipal solid waste each year. This separates recycled waste from what can be burned, so nothing goes to landfill. The remaining waste goes to an incineration plant producing electricity for 25 000 homes.

The second successful policy has been to increase recycling volumes from 50 kg per person per year in 2004–05 to 200 kg per person per year in 2017–18. Half of all waste is now recycled in Bristol, again reducing landfill. The city council collects different materials for recycling at the kerbside to make it easy to recycle. This is reinforced in schools which have recycling policies in the curriculum.

Evidence – uses evidence of data to support the point

Point – the second point is how Bristol has increased the amount of recycling

Explanation – extends the point using kerbside collection and education to reinforce the point

Point – the third point is giving recycling and waste companies targets

The final policy encourages recycling and waste companies to increase what they collect from people's houses by using targets. Bristol City Council gives specialised targets for particular materials (e.g. composted waste, plastics, or bottles) so this <u>increases their motivation</u> to collect all the recycling they can. This puts responsibility on to companies when they get the contract to collect waste and <u>has led to increases in all types of recycling</u>.

Evidence – uses evidence to show how Bristol uses targets

Explanation – extends the explanation to show how targets work for waste companies

Links – there are nine underlined links. All the phrases used (e.g. 'successful') help to indicate how the policies succeeded. These are all evaluative links, boosting the AO2 elements of the answer.

 Examiner feedback

This is a strong answer. The descriptor for Level 3 applies to this answer as follows:

For AO1:

- *There is evidence of detailed knowledge, of clear understanding of concepts, and of physical and human processes throughout the answer.*

For AO2:

- *Applies knowledge and understanding of geographical information/ideas logically, making relevant connections/relationships.*
- *Applies knowledge and understanding of geographical information/ideas to produce a full and coherent interpretation that is relevant and supported by evidence.*
- *Applies knowledge and understanding of geographical information/ideas to make supported judgements about the significance of factors throughout the response, leading to a balanced and coherent argument.*

The underlined links back to the question demonstrate that the candidate can 'evaluate'. By meeting the Level 3 descriptors fully, the answer earns all 9 marks.

In this section you'll learn how to...

- maximise marks on 9-mark questions on Population and the environment, using the command term 'How far do you agree?'.

Tackling 9-mark questions about Population and the environment

This topic is assessed on Paper 2, Section C, Question 4.

- Before beginning this section, read section 1.6, which will tell you about the skills you need to answer 9-mark questions, and the mark scheme.

Try this 9-mark question on Population and the environment

> This question assesses AO1(knowledge and understanding) for 4 marks and AO2 (application) for 5 marks.

Question

How far do you agree with the statement that deteriorating soil quality is due to mismanagement rather than natural causes? (9 marks)

1 Plan your answer

Before attempting to answer the question, remember to **BUG** it. That means:

✓ **Box** the command word.
✓ **Underline** the following:
  - the theme
  - the focus
  - any evidence required
  - the number of points needed.
✓ **Glance** back over the question – to make sure you include everything in your answer.

Use the **BUG** on the next page to plan your own answer.

Five steps to success!

Five steps to help you write top quality answers

The following five steps are used in this book to help you get the best marks.

1. **Plan your answer** – decide what to include and how to structure your answer.

2. **Write your answer** – use the answer spaces to write your answer.

3. **Mark your answer** – use the mark scheme (section 1.6) to self- or peer-mark your answer. You can also use this to assess sample answers in step 4 below.

4. **Sample answers** – two sample answers are given to show you how to maximise marks for a question.

5. **Marked sample answer** – this is the same answer that you used for step 4 above, but is marked and annotated, so that you can compare with your own answers.

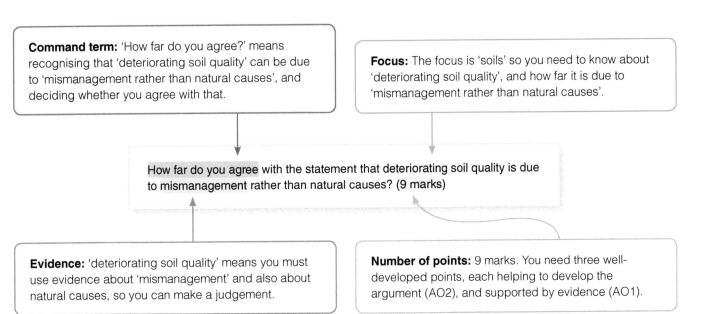

Command term: 'How far do you agree?' means recognising that 'deteriorating soil quality' can be due to 'mismanagement rather than natural causes', and deciding whether you agree with that.

Focus: The focus is 'soils' so you need to know about 'deteriorating soil quality', and how far it is due to 'mismanagement rather than natural causes'.

How far do you agree with the statement that deteriorating soil quality is due to mismanagement rather than natural causes? (9 marks)

Evidence: 'deteriorating soil quality' means you must use evidence about 'mismanagement' and also about natural causes, so you can make a judgement.

Number of points: 9 marks. You need three well-developed points, each helping to develop the argument (AO2), and supported by evidence (AO1).

PEEL your answer

Use PEEL notes to structure your answer. This will help to explain your ideas to the examiner clearly. PEEL has four stages:

- **P**oint – Make three well-developed points for 9 marks. These are the **AO2** parts of the answer where you argue how far you agree with the statement that deteriorating soil quality is due to mismanagement rather than natural causes.
- **E**vidence – Include details about specific ways in which soils can deteriorate to illustrate your argument. It's important to quote examples, such as salinisation. This is **AO1** – using knowledge to support your argument.
- **E**xplain – Give a reason for each point. Use starter phrases such as: '*The problem of soil erosion is due to natural causes because ...*'. This extends your **AO2** ability because you'll explain how soil erosion problems have arisen.
- **L**ink back to the question – Use the wording of the question in your answer to show you recognise what the question is asking, e.g. '*deteriorating soil quality is due to mismanagement because ...*'.

 Tip

Remember, quality not quantity

You will not be marked on the **number** of points you make, but on the **quality** of your answer. That means the quality of the argument (AO2), the evidence you give (AO1), and how you link back to the question.

Remember the AOs!

It's important to know the AOs for this question.
- **AO1** (4 marks) is about showing knowledge and understanding of specific ways in which soils are deteriorating. Use brief examples – not long case studies.
- **AO2** (5 marks) is about using evidence to judge whether these are due to mismanagement rather than natural causes.

② Write your answer

> How far do you agree with the statement that deteriorating soil quality is due to mismanagement rather than natural causes? (9 marks)
>
> _____
>
> _____
>
> _____
>
> _____
>
> _____
>
> _____
>
> _____
>
> _____
>
> _____
>
> _____
>
> _____
>
> _____
>
> _____
>
> _____
>
> _____
>
> _____
>
> _____
>
> *You can continue on file paper if you need more space*

③ Mark your answer

1. To help you to identify well-structured points, highlight or underline:

- points in red
- explanations in orange
- evidence in blue
- links back to the question by underlining.

2. Use the mark scheme in section 1.6. Remember that:

- 9-mark questions are marked by choosing a level and a mark within it, based upon the answer as a whole.
- High-scoring answers must include **both** AOs. One which includes only AO1 without explaining the answer for AO2 can only gain a maximum of 4 marks.

Things to watch out for

Study Sample Answer 1 below.

- The answer has strengths, so identify them.
- Suggest what prevents it from getting a higher mark.

Clues:
Read how the candidate demonstrates:
a) **AO1** – what does the candidate know about deteriorating soil quality?
b) **AO2** – how well does the candidate judge the extent to which these are the result of mismanagement rather than natural causes?

Sample Answer 1

Soil is being lost at between 10 and 40 times the rate at which it's replaced. 40% of world soils are in a poor state, caused by problems like soil erosion. Soil erosion occurs after deforestation when soil is deprived of protection by tree canopies and of being held together by roots. Heavy rainfall, especially in tropical monsoon seasons, causes soil erosion. Erosion can also be caused by cattle overgrazing, exposing the soil to drought and wind, so the fertile top soil is blown away.

Sometimes soil erosion occurs when hedgerows are removed. It helps farmers by enlarging fields which makes them easier to plough or harvest and increases the area that can be cultivated. But it also means that there are no hedgerows to break the strength of wind or rain during storms, so soils are easily eroded.

Other soil problems include waterlogging, which means that seedlings cannot develop roots and respire properly when the soil lacks air spaces. Sometimes this is caused by soil compaction due to machinery overuse in a field, and sometimes it is due to high water tables when water is unable to infiltrate after a very wet period. Salinisation is another problem, where waterlogging brings soil minerals to the surface, which increase in concentration and prevent plant growth.

	Remember! Check for both AO1 (knowledge and understanding) and AO2 (application).		
Strengths of the answer			
Ways to improve the answer			
Level		**Mark**	

Examiner feedback

The answer has strengths.

- The candidate shows knowledge and understanding of different soil problems such as erosion or waterlogging. (AO1)
- However, the candidate does not evaluate these problems (AO2). It is clear that processes such as soil erosion ARE caused by human action, e.g. deforestation, but the candidate doesn't actually say how important this is, even though the evidence is there.
- To earn a higher mark, the candidate needs to explain how far the causes are mismanagement rather than natural. (AO2)

The answer fits criteria for different levels. For AO1, the candidate fits the criterion for Level 2:

- There is evidence of *clear and relevant knowledge,* of a *sound understanding of concepts,* and *grasp of physical and processes.* Evidence of these is sustained though perhaps *with minor inaccuracy.*

However, AO2 is only Level 1–2:

- *Knowledge and understanding* is rarely applied to the question. (Level 1)
- *Links between different causes* are not expanded upon. (Level 1)
- There is evidence of *general analysis, supported with some evidence.* (Level 2)
- There is little *evaluation of the causes of soil deterioration.* (Level 1)

The answer therefore gets a 'best fit' mark of Level 2, with 4 marks. It's as though the candidate has paid little attention to answering the question.

 Now mark this one!

 Question recap

Read through Sample Answer 2 below.

- a) Go through the answer using the three colours in section 3, and underline any links back to the question.
- b) Look for evidence of both knowledge and understanding.
- c) Use the mark scheme in section 1.6 to decide on a level and mark.

How far do you agree with the statement that deteriorating soil quality is due to mismanagement rather than natural causes? (9 marks)

Sample Answer 2

The key causes of soil deterioration include soil erosion, waterlogging and salinisation. Many of these causes are natural but are worsened by mismanagement, like the decision to undertake deforestation, especially in rainforests which removes the protection of tree canopies and roots. Heavy rains cause soil erosion, which is natural, but much faster than before deforestation. Erosion is also caused by drought, e.g. the American Midwest where ploughing up natural grassland exposed soil to drought and wind, so fertile top soil was eroded.

Hedgerow removal also results from mismanagement, and causes soil erosion. Farmers enlarge fields to make it easier to mechanise and increase the area under cultivation. But lack of hedgerows removes windbreaks, so soils are easily eroded. This has occurred for two centuries in the Fens in the UK.

Other soil problems include waterlogging, when soil lacks air spaces, and seedlings cannot develop roots or respire properly. Sometimes this is caused by mismanagement when farm machinery repeatedly compacts soil in a field. There are also natural causes, such as in the winter of 2014 in the UK when high water tables resulted from a very wet winter which saturated soils. But sometimes waterlogging causes salinisation, when it brings minerals to the surface, increasing their concentration and preventing root absorption and growth. This is due to human mismanagement as it occurs in areas like the Colorado valley or in Australia where overirrigation has caused waterlogging and salinisation.

	Remember! Check for both AO1 (knowledge and understanding) and AO2 (application).		
Strengths of the answer			
Ways to improve the answer			
Level		**Mark**	

⑤ Marked sample answers

Sample Answer 2 is marked below. The text has been highlighted to show how well each answer has structured points.

The following have been highlighted:

- points in red
- explanations in orange
- evidence in blue
- <u>underlined</u> points are where the candidate links to the question concerning natural causes versus mismanagement.

Marked sample answer 2

Explanation – extends the point by explaining the impact of heavy rains

Point – the second point is about hedgerow removal

Evidence – uses evidence from the Fens to support the point

> The key causes of soil deterioration include soil erosion, waterlogging and salinisation. Many of these causes are natural but are worsened by mismanagement, like the decision to undertake deforestation, especially in rainforests which removes the protection of tree canopies and roots. Heavy rains cause soil erosion, which is natural, but much faster than before deforestation. Erosion is also caused by drought, e.g. the American Midwest where ploughing up natural grassland exposed soil to drought and wind, so fertile top soil was eroded.
>
> Hedgerow removal also results from mismanagement, and causes soil erosion. Farmers enlarge fields to make it easier to mechanise and increase the area under cultivation. But lack of hedgerows removes windbreaks, so soils are easily eroded. This has occurred for two centuries in the Fens in the UK.

Point – the first point is about deforestation, linked to mismanagement

Evidence – uses evidence of the American Midwest to highlight the causes of erosion

Explanation – extends the point by explaining the reasons but also highlighting the risks

Point – the third point is about waterlogging

Evidence – uses evidence of natural causes in 2014

Other soil problems include waterlogging, when soil lacks air spaces, and seedlings cannot develop roots or respire properly. Sometimes this is caused by mismanagement when farm machinery repeatedly compacts soil in a field. There are also natural causes such as in the winter of 2014 in the UK when high water tables resulted from a very wet winter which saturated soils. But sometimes waterlogging causes salinisation, when it brings minerals to the surface, increasing their concentration and preventing root absorption and growth. This is due to human mismanagement as it occurs in areas like the Colorado valley or in Australia where overirrigation has caused waterlogging and salinisation.

Explanation – extends the point by explaining how waterlogging is caused

Point – the final point links waterlogging to desalinisation

Evidence – uses examples from the Colorado valley and Australia

Explanation – extends the explanation to show how irrigation has had impacts on the soil

Links – there are six underlined links. All the phrases used identify whether the cause is natural or mismanagement. These all help to answer the question, boosting the AO2 elements of the answer.

✓ Examiner feedback

This is a strong answer. The descriptor for Level 3 applies to this answer as follows:

For AO1:

- *There is evidence of detailed knowledge, of clear understanding of concepts, and of physical and human processes throughout the answer.*

For AO2:

- *Applies knowledge and understanding of geographical information/ideas logically, making relevant connections/relationships.*
- *Applies knowledge and understanding of geographical information/ideas to produce a full and coherent interpretation that is relevant and supported by evidence.*
- *Applies knowledge and understanding of geographical information/ideas to make supported judgements about the significance of factors throughout the response, leading to a balanced and coherent argument.*

The underlined links back to the question all demonstrate that the candidate can make a judgement about the statement in the question. By meeting the Level 3 descriptors fully, the answer earns all 9 marks.

In this section you'll learn how to...

- maximise marks on 9-mark questions on Resource security, using the command term 'Analyse'.

Tackling 9-mark questions about Resource security

This topic is assessed on Paper 2, Section C, Question 5.

- Before beginning this section, read section 1.6, which will tell you about the skills you need to answer 9-mark questions, and the mark scheme.

Try this 9-mark question on Resource security

> This question assesses AO1 (knowledge and understanding) for 4 marks and AO2 (application) for 5 marks.

Question

Analyse the reasons why the movement of energy resources between their source areas and where they will be consumed can be prone to risk. (9 marks)

1 Plan your answer

Before attempting to answer the question, remember to **BUG** it. That means:

✓ **Box** the command word.
✓ **Underline** the following:
  - the theme
  - the focus
  - any evidence required
  - the number of points needed.
✓ **Glance** back over the question – to make sure you include everything in your answer.

Use the **BUG** on the next page to plan your own answer.

 **Five steps to success!**

Five steps to help you write top quality answers

The following five steps are used in this book to help you get the best marks.

1. **Plan your answer** – decide what to include and how to structure your answer.

2. **Write your answer** – use the answer spaces to write your answer.

3. **Mark your answer** – use the mark scheme (section 1.6) to self- or peer-mark your answer. You can also use this to assess sample answers in step 4 below.

4. **Sample answers** – two sample answers are given to show you how to maximise marks for a question.

5. **Marked sample answer** – this is the same answer that you used for step 4 above, but is marked and annotated, so that you can compare with your own answers.

Command term: 'Analyse' means to break something down into individual components and say how each contributes to the theme/topic. This means you must identify the kinds of reasons which pose risks to energy transport, e.g. political or environmental.

Focus: The focus is 'energy' so you need to know about how energy resources are transported, and the risks to which these movements are subjected.

Analyse the reasons why the movement of energy resources between their source areas and where they will be consumed can be prone to risk. (9 marks)

Evidence: 'prone to risk' means you must use evidence about where the movement of energy resources faces risks, and identify the kinds of risks it faces.

Number of points: 9 marks. You need three well-developed points, each helping to develop the argument (AO2), and supported by evidence (AO1).

PEEL your answer

Use PEEL notes to structure your answer. This will help to explain your ideas to the examiner clearly. PEEL has four stages:

- **P**oint – Make three well-developed points for 9 marks. These are the **AO2** parts of the answer where you explain how the movement of energy resources between source areas and where they will be consumed can be prone to risk.
- **E**vidence – Include details about specific risks to movement of energy resources e.g. piracy. This is **AO1** – using knowledge to support your argument.
- **E**xplain – Give a reason for each point. Use starter phrases such as: '*The risks are often due to conflicts such as ...*'. This extends your **AO2** ability because you'll explain how risks have arisen.
- **L**ink back to the question – Use the wording of the question in your answer to show you recognise what the question is asking, e.g. '*the risks to movement of energy resources are often due to political instability because ...*'.

 Tip

Remember, quality not quantity

You will not be marked on the **number** of points you make, but on the **quality** of your answer. That means the quality of the argument (AO2), the evidence you give (AO1), and how you link back to the question.

Remember the AOs!

It's important to know the AOs for this question.
- **AO1** (4 marks) is about showing knowledge and understanding of specific risks to which energy movements can be subjected. Use brief examples – not long case studies.
- **AO2** (5 marks) is about using evidence to suggest which of these risks are most important, or whether they are of a particular type.

2 Write your answer

Analyse the reasons why the movement of energy resources between their source areas and where they will be consumed can be prone to risk. (9 marks)

_____

_____

_____

_____

_____

_____

_____

_____

_____

_____

_____

_____

_____

_____

_____

_____

_____

_____

_____

_____

You can continue on file paper if you need more space

3 Mark your answer

1. To help you to identify well-structured points, highlight or underline:

- points in red
- explanations in orange
- evidence in blue
- links back to the question by underlining.

2. Use the mark scheme in section 1.6. Remember that:

- 9-mark questions are marked by choosing a level and a mark within it, based upon the answer as a whole.
- High-scoring answers must include **both** AOs. One which includes only AO1 without explaining the answer for AO2 can only gain a maximum of 4 marks.

Things to watch out for

Study Sample Answer 1 below.

• The answer has strengths, so identify them.
• Suggest what prevents it from getting a higher mark.

Clues:
Read how the candidate demonstrates:
a) **AO1** – what does the candidate know about energy movement and the risks faced?
b) **AO2** – how well does the candidate judge which of these risks are most important, or whether they are of a particular type?

Sample Answer 1

The movement of energy resources between their source areas and where they will be consumed can be prone to risk. This can depend upon whether they are fossil fuels, or more sensitive ones, such as uranium for the nuclear industry.

There are many ways of transporting energy resources from their source areas. They include pipelines for oil and gas, such as the pipelines between Russia's oil and gas fields and Europe where these are consumed. This is risky as pipelines might get destroyed or leak.

Electricity is sent via transmission lines, such as the electricity that the UK imports from France at peak times. This can be risky as storms might blow the lines down. Shipping is another way of moving energy resources, such as oil and gas, or other fossil fuels, such as coal. These can also be moved by road and rail if they are being moved less far. Shipping can be risky as a ship might run aground and cause a pollution incident.

Sometimes the movement of energy resources faces risks when countries are in conflict, such as Russia and Ukraine. This could include terrorists who wish to get hold of the resource. It would mean disrupting supplies.

	Remember! Check for both AO1 (knowledge and understanding) and AO2 (application).		
Strengths of the answer			
Ways to improve the answer			
Level		**Mark**	

Examiner feedback

The answer has strengths.

- The candidate shows knowledge and understanding of different energy movements such as oil or electricity, and brief details about some risks. (AO1)
- However, the candidate does not analyse these risks (AO2). They are not spelled out in much detail, nor expanded upon, or categorised into types.
- To earn a higher mark, the candidate needs to explain the risks in slightly more detail, but also to say what types of risks these are e.g. political, or to the environment. (AO2)

For AO1, the answer fits the criteria for Level 2:

- There is evidence of *clear and relevant knowledge,* of a *sound understanding of concepts,* and *grasp of physical and human processes.* Evidence of these is sustained though perhaps *with minor inaccuracy.*

However, AO2 fits the criteria more for Level 1:

- *Knowledge and understanding* is applied in limited ways.
- *Different concepts or links between different parts of the answer* are basic, with limited development or expansion.
- There is evidence of basic *analysis and evaluation,* supported with *limited evidence.*

The answer therefore gets a 'best fit' mark of low Level 2, with 4 marks.

4 Now mark this one!

Read through Sample Answer 2 below.

a) Go through the answer using the three colours in section 3, and underline any links back to the question.
b) Look for evidence of both knowledge and understanding.
c) Use the mark scheme in section 1.6 to decide on a level and mark.

 Question recap

Analyse the reasons why the movement of energy resources between their source areas and where they will be consumed can be prone to risk. (9 marks)

Sample Answer 2

Energy movements between source areas and where energy is consumed can be subject to several risk factors, all of which are political. The first is internal political tensions within countries through which energy resources must travel. An example is the gas pipeline between Azerbaijan and Europe because ongoing tensions within Azerbaijan and Armenia make supplies subject to disruption, with risks of terrorism from ISIL. This can disrupt supplies by delaying shipments, and can affect prices.

Similar political tensions have built up between rival groups in the Middle East, where shipping routes pass through the Straits of Hormuz. These are prone to risk of disruption because of tensions between Shia and Sunni Muslims who control half of the straits each. Similarly, pirates operating from within failed states, such as Somalia and Yemen, disrupt shipping passing into the Red Sea, because neither country has law enforcement resources to control increased piracy. Pirates operating in the region have held ships and their crews hostage subject to ransom.

The third risk factor is purely political and concerns political relations between states, such as those between Russia, Ukraine, and countries of the EU. Russia has some of the world's largest gas reserves, yet Ukraine is crucial to transporting the gas because pipelines must cross its territory to access European markets. In 2004 (and several times since) Russia has cut supplies or has quadrupled prices to Ukraine, all of which have affected reliability of European supplies and price levels.

	Remember! Check for both AO1 (knowledge and understanding) and AO2 (application).
Strengths of the answer	
Ways to improve the answer	
Level	**Mark**

5 Marked sample answers

Sample Answer 2 is marked below. The text has been highlighted to show how well each answer has structured points.

The following have been highlighted:

- points in red
- explanations in orange
- evidence in blue
- underlined points are where the candidate links to the question concerning risks to the movement of energy resources.

Marked sample answer 2

Point – the first point identifies internal political tensions

Explanation – extends the point by explaining the impact on delay and costs

Point – the second point identifies tensions between rival groups

Energy movements between source areas and where energy is consumed can be subject to several risk factors, all of which are political. The first is internal political tensions within countries through which energy resources must travel. An example is the gas pipeline between Azerbaijan and Europe because ongoing tensions within Azerbaijan and Armenia make supplies subject to disruption, with risks of terrorism from ISIL. This can disrupt supplies by delaying shipments, and can affect prices.

Similar political tensions have built up between rival groups in the Middle East, where shipping routes pass through the Straits of Hormuz. These are prone to risk of disruption because of tensions between Shia and Sunni Muslims who control half of the straits each. Similarly, pirates operating from within failed states, such as Somalia and Yemen, disrupt shipping passing into the Red Sea, because neither country has law enforcement resources to control increased piracy. Pirates operating in the region have held ships and their crews hostage subject to ransom.

Evidence – uses evidence of tensions within Azerbaijan and Armenia

Explanation – extends the point by explaining tensions between Shia and Sunni Muslims

Evidence – uses evidence from Somalia and Yemen

Explanation – extends the point by explaining tensions between Russia and Ukraine

Point – the third point identifies political relations between states as a factor

> The third risk factor is purely political and concerns <u>political relations between states</u>, such as those between Russia, Ukraine, and countries of the EU. Russia has some of the world's largest gas reserves, yet Ukraine is crucial to transporting the gas because pipelines must cross its territory to access European markets. In 2004 (and several times since) Russia has cut supplies or has quadrupled prices to Ukraine, all of which have affected reliability of European supplies and price levels.

Evidence – uses evidence of events in 2004 to explain impacts on gas supplies and prices in Europe

Links – there are four underlined links. All the phrases used identify the evidence that the candidate is analysing the different risks by spelling out what type they are. These boost the AO2 elements of the answer.

Examiner feedback

This is a strong answer – there are three well-developed points. The descriptor for Level 3 applies to this answer as follows:

For AO1:

- *There is evidence of detailed knowledge, of clear understanding of concepts, and of physical and human processes throughout the answer.*

For AO2:

- *Applies knowledge and understanding of geographical information/ideas logically, making relevant connections/relationships.*
- *Applies knowledge and understanding of geographical information/ideas to produce a full and coherent interpretation that is relevant and supported by evidence.*
- *Applies knowledge and understanding of geographical information/ideas to make supported judgements about the significance of factors throughout the response, leading to a balanced and coherent argument.*

The underlined links back to the question all demonstrate that the candidate is analysing.
By meeting the Level 3 descriptors fully, the answer earns all 9 marks.

In this section you'll learn how to...

- maximise marks on 20-mark questions on the Carbon cycle, which use the command word 'Assess'.

Tackling 20-mark questions about the Carbon cycle

This topic is assessed on Paper 1, Section A, Question 1.

- Before beginning this section, you should study section 1.7 carefully which will tell you about the skills you need to answer 20-mark questions, and the mark scheme.

Try this 20-mark question on the Carbon cycle

> This question assesses AO1 for 10 marks and AO2 for 10 marks.

Question

Assess the statement that 'the most important factor modifying the carbon cycle at present is the global demand for energy'. (20 marks)

① Plan your answer

Before attempting to answer the question, remember to **BUG** it. That means:

✓ **Box** the command word.
✓ **Underline** the following:
  - the theme
  - the focus
  - any evidence required
  - the number of points needed.
✓ **Glance** back over the question – to make sure you include everything in your answer.

Use the **BUG** on the next page to plan your own answer.

Five steps to success!

Five steps to help you write top quality answers

The following five steps are used in this book to help you get the best marks.

1. **Plan your answer** – decide what to include and how to structure your answer.

2. **Write your answer** – use the answer spaces to write your answer.

3. **Mark your answer** – use the mark scheme (section 1.7) to self- or peer-mark your answer. You can also use this to assess sample answers in step 4 below.

4. **Sample answers** – two sample answers are given to show you how to maximise marks for a question.

5. **Marked sample answer** – this is the same answer that you used for step 4 above, but is marked and annotated, so that you can compare with your own answers.

Command word: 'Assess' means use evidence to determine and weigh up the extent to which global energy demands are the most important factor modifying the carbon cycle at present.

Focus: The focus is about 'the global demand for energy' and how far this is altering the carbon cycle, compared to other factors such as land-use changes, or deforestation, for example.

Assess the statement that 'the most important factor modifying the carbon cycle at present is the global demand for energy'. (20 marks)

Evidence: How the global demand for energy is altering the carbon cycle, from examples you have studied, and how far any other factors are important

No of points: 20 marks. You need five or six reasoned points, preferably balanced across both sides of the argument (AO2), supported by evidence (AO1).

PEEL your answer

Use PEEL notes to structure your answer. This will help to explain your ideas to the examiner clearly. PEEL has four stages:

- **P**oint – Make five or six developed points for 20 marks. These are the **AO2** parts of the answer where you discuss how far the global demand for energy is altering the carbon cycle, and show how far this is significant when compared to other changes.
- **E**vidence – Include details about specific ways in which demands for energy and burning fossil fuels are altering the carbon cycle in order to to illustrate your argument. It's important to use examples, such as whether they are short- or long-term. This is **AO1** – using knowledge to support your argument.
- **E**xplain – Give a reason for each point. Use starter phrases such as: '*This change to the carbon cycle is important because ...*'. This extends your **AO2** ability because you'll show that you can identify the significance of particular changes.
- **L**ink back to the question – Use the wording of the question in your answer to show you recognise what it is asking, e.g. '*Global energy demand has brought about considerable change because ...*'.

 Tip

Remember, quality not quantity

You will not be marked on the **number** of points you make, but on the **quality** of your answer. That means the quality of the argument (AO2), the evidence you give (AO1), and how you link back to the question.

Remember the AOs!

It's important to know the AOs for this question.
- **AO1** (10 marks) is about demonstrating knowledge and understanding of ways in which global energy use changes the carbon cycle. Use brief examples – not long case studies.
- **AO2** (10 marks) is about using evidence to assess how far global energy demand is the main factor causing these changes, and how it compares to other factors.

2 Write your answer

Using file paper, write your answer to the following question.

Question

Assess the statement that 'the most important factor modifying the carbon cycle at present is the global demand for energy'. (20 marks)

3 Mark your answer

1. To help you to identify well-structured points, highlight or underline:

- points in red
- explanations in orange
- evidence in blue
- links back to the question by underlining.

Look for an introduction and conclusion, which are essential for questions with the command word 'Assess'.

2. Use the mark scheme in section 1.7. Remember that:

- 20-mark questions are marked by choosing a level and a mark within it, based upon the answer as a whole
- a high-scoring answer must include **both** AOs. One which includes only AO1 without applying the question for AO2 can only gain a maximum of 10 marks.

Things to watch out for

Study Sample Answer 1 below.

- The candidate has done some things well, so identify which qualities are good.
- Identify what prevents the answer from getting a higher mark.

  Clues:
  Look at how the candidate demonstrates:
  a) **AO1** – what does the candidate know about changes to the carbon cycle and the part played by global energy demand in bringing about these changes?
  b) **AO2** – how well does the candidate develop arguments about how important the role of energy demand is in causing these changes? Are any other causes also identified as important?

Remember the introduction and conclusion!

All 20-mark questions with the command word 'Assess' need to have:
- a brief (1–2 sentence) introduction
- a final short paragraph conclusion.

Sample Answer 1

The carbon cycle means the circulation of carbon in the biosphere. When this cycle occurs, plants convert CO_2 in the atmosphere into organic compounds by photosynthesis. As plants and animals respire, they use oxygen and then release CO_2 back into the atmosphere. This also happens when fossil fuels are combusted. This can change the cycle because when fossil fuels are burned, more CO_2 is returned to the atmosphere than before, as the rocks stored carbon. This means that climate is likely to change.

The other thing that can change the carbon cycle is deforestation. This removes plants which would absorb carbon from the atmosphere and that too leads to increased carbon dioxide in the air. A lot of this is driven by demands for energy in the world as demand rises with more cars and more countries industrialising.

CO_2 is a greenhouse gas, which have increased in the atmosphere since the 18th century due to industrialisation. The other gases include methane. But the pace is picking up because, since the 1980s, most global CO_2 emissions have come from burning fossil fuels. Climate scientists think that this increases global temperatures, because of extra greenhouse gases in the atmosphere. This is called an enhanced greenhouse effect. As global temperatures increase, the level of water vapour in the atmosphere increases; there is greater evaporation and condensation, increasing cloud cover, which traps heat in the atmosphere and increases temperatures.

Greenhouse gases help to maintain the Earth's temperature naturally, which also affects global distributions of temperature and precipitation – so changing the amounts in the atmosphere is likely to alter these. One cause is growing energy demand. Growing use of fossil fuel energy creates more greenhouse gases, which impacts on carbon emissions. These change the carbon cycle.

Other human factors also modify the carbon cycle, especially land-use changes causing deforestation and changes to oceans, which also disrupt the carbon cycle. Phytoplankton in oceans sequester CO_2 by photosynthesis which makes them transfer it out of the atmosphere and into ocean carbon stores. Photosynthesis on land enables plants to sequester CO_2 which is released into the atmosphere through respiration and decomposition. Anything that affects the level of phytoplankton in the oceans or the forested area will affect the level of carbon sequestration, and therefore the atmosphere.

The Earth has several carbon reservoirs, which are sources that add carbon to the atmosphere, as well as carbon sinks which remove it. When these sources and sinks are equal, the carbon cycle is in a balance. When that happens, a steady amount of CO_2 in the atmosphere helps to keep global temperatures steady. But human activities increase CO_2 inputs, and this can happen without increasing natural sinks such as oceans or forests. So this increases atmospheric stores of carbon and increases global temperatures. Fossil fuel combustion changes the balance of carbon flows and stores – and as more carbon is released from stores, then the flows are increased causing climate change.

Changes in climate are likely to vary. Weather scientists think that extreme weather will increase with bigger storms and droughts. The Arctic is worst affected globally but other areas can expect more drought, like sub-Saharan Africa and Australia if El Niño is affected.

To conclude, the most important factor modifying the carbon cycle is the world's insatiable demand for energy which is causing more CO_2 to be emitted and change the carbon cycle.

	Remember! Check for both AO1 (knowledge and understanding) and AO2 (application).	
Strengths of the answer		
Ways to improve the answer		
Level		Mark	

✓ **Examiner feedback**

As with all 20-mark questions, there are four levels for this question. The candidate has made some relevant points.

- The candidate has good knowledge and understanding (AO1) of how the carbon cycle works, with words such as sequestration, stores and flows showing that s/he knows the components of the cycle.
- There is some application (AO2) – the candidate refers to changes in the atmosphere caused by increased carbon emissions and how these affect the cycle.
- However, there is no detail about how other factors (e.g. land-use change) may be affecting the cycle.
- To earn a higher mark, the candidate needs less about how the carbon cycle works in the atmosphere and more about causes. Land-use change is barely considered, yet this is an important cause, and more is needed about why global energy demands are rising. The conclusion is weak.

For AO1, the descriptor for Level 4 applies:

- *Knowledge and understanding* is clear, accurate and relevant to the argument, in explaining concepts and processes, and in explaining geographical scale and change over time.

For AO2, Level 3 of the mark scheme applies to the first part:

- *Application is sound in different contexts (e.g. in different examples used), shows thorough, clear, and relevant analysis* – the argument is never established how energy demands actually affect the carbon cycle. The conclusion is weak – and this part of the answer is possibly only Level 2.

The answer is a mix of Levels 3 and 4. On a 'best fit', this results in low-mid Level 3, i.e. 13 marks.

④ Now mark this one!

Read through Sample Answer 2 below.

  a) Go through the answer using the three colours in section 3, and underline any links back to the question.
  b) Look for evidence of both knowledge and understanding.
  c) Use the mark scheme in section 1.7 to decide which level it is in and how many marks it is worth.

 Question recap

Assess the statement that 'the most important factor modifying the carbon cycle at present is the global demand for energy'. (20 marks)

Sample Answer 2

The carbon cycle is the circulation of carbon in the biosphere, by which CO_2 in the atmosphere converts into organic compounds by plants, which, following consumption by other organisms, are returned to the atmosphere as CO_2 by respiration, decay and combustion of fossil fuels. The cycle is easily modified by human intervention such as by burning fossil fuels or by deforestation. The more fossil fuels are consumed, the greater the disturbance of the carbon cycle by human activity.

The growth in demand for energy between the first industrial revolution in the 18th century, and, more recently, by emerging and high income countries in the 21st century has led to significant changes in CO_2 emissions. Growing demand for energy is highly significant in modifying the carbon cycle and results from global economic growth, which remains a priority for most governments, and for intergovernmental organisations such as the G7 or G20. Fossil fuels now dominate the global economy and there is little sign of that changing, despite some growth of renewables. Increasing demand is therefore likely to continue.

Globally, oil is the fuel of choice for transport systems, though coal remains the most important primary source of energy for electricity production. At one stage in the early 21st century, China was opening a coal-fired power station every three weeks. Reductions in energy prices in recent years (led by oil) have ensured that demand continues to grow, with little prospect of it decreasing. The pace of growth in fossil fuels has changed so that China is now the largest producer globally of renewable energy technology, from wind generators to solar panels, but fossil fuels still predominate. Most emerging countries, such as India, use more fossil fuels every year as their economies grow. It is the growing use of fossil fuel energies that has increased greenhouse gases dramatically.

Not only does energy consumption increase with economic growth, but even obtaining energy is in itself responsible for increased greenhouse gas emissions. Many methods of extracting unconventional energy sources generate high levels of carbon emissions, such as tar sands, oil shale and fracking. 'Peak oil' is less of a concern now that unconventional sources have made such an impact in reducing oil process globally; the general trend of global oil prices since 2008 has been downward. US oil prices have fallen sharply since 2012, due to huge new supplies of oil and shale gas from the USA and Canada. To compete, OPEC cut its prices to maintain its market share.

Most assessments, for example by the IPCC, identify the burning of fossil fuels as the major cause of anthropogenic climate change and the ways in which it impacts on other systems, such as ocean temperatures. Earth's carbon reservoirs act as sources and sinks. If sources and sinks are equal, the carbon cycle is in equilibrium. However, human activities have led to increased CO_2 emissions, without corresponding increases in natural sinks (e.g. oceans and forests). Increasing atmospheric stores of carbon are widely believed to be the main cause of rising global temperatures, and fossil fuel combustion has altered the balance of these carbon pathways and stores. With carbon being released in larger amounts by burning, flows have greatly increased while stores on Earth have decreased.

Burning fossil fuels has important consequences for the working of the carbon cycle, but land-use changes also disrupt the way the cycle works. Land-use change, such as deforestation for palm oil or for mining, causes major changes to the chemistry and temperature of oceans, which may disrupt their role as stores in the carbon cycle. Other factors, such as peat extraction and land-use changes (e.g. forest clearance for cattle farming) are also significant. Deforestation has increased in rainforests especially, followed by the proliferation of palm oil plantations for biofuel production. Along with rainforest clearance, increased intensification of farming has also led to land-use changes, which impact on energy usage.

So the statement that the most important factor in modifying the carbon cycle is the world's growing demand for energy is true, and is the main factor driving changes to the carbon cycle, though land-use changes are also significant.

	Remember! Check for both AO1 (knowledge and understanding) and AO2 (application).		
Strengths of the answer			
Ways to improve the answer			
Level		**Mark**	

⑤ Marked sample answers

Sample Answer 2 is marked below. The text has been highlighted to show how well each answer has structured points. The following have been highlighted:

- points in red
- explanations in orange
- evidence in blue
- underlined points are where the candidate links to the question to identify the importance of global demand for energy in modifying the carbon cycle.

Look for the introduction and conclusion, which are essential for questions with the command word 'Assess'.

Marked sample answer 2

Link – the candidate introduces the answer well, with a strong definition of the carbon cycle, as well as setting out the context

The carbon cycle is the circulation of carbon in the biosphere, by which CO_2 in the atmosphere converts into organic compounds by plants, which, following consumption by other organisms, are returned to the atmosphere as CO_2 by respiration, decay and combustion of fossil fuels. The cycle is easily modified by human intervention such as by burning fossil fuels or by deforestation. The more fossil fuels are consumed, the greater the disturbance of the carbon cycle by human activity.

Explanation – expands the point referring to economic growth as a global aim

The growth in demand for energy between the first industrial revolution in the 18th century, and, more recently, by emerging and high income countries in the 21st century has led to significant changes in CO_2 emissions. Growing demand for energy is highly significant in modifying the carbon cycle and results from global economic growth, which remains a priority for most governments, and for intergovernmental organisations such as the G7 or G20. Fossil fuels now dominate the global economy and there is little sign of that changing, despite some growth of renewables. Increasing demand is therefore likely to continue.

Point – links the increase in CO_2 emissions to industrialisation

Evidence – uses evidence of the G7 and G20

Point – refers to coal and oil as the main drivers of transport and electricity

Globally, oil is the fuel of choice for transport systems, though coal remains the most important primary source of energy for electricity production. At one stage in the early 21st century, China was opening a coal-fired power station every three weeks. Reductions in energy prices in recent years (led by oil) have ensured that demand continues to grow, with little prospect of it decreasing. The pace of growth in fossil fuels has changed so that China is now the largest producer globally of renewable energy technology, from wind generators to solar panels, but fossil fuels still predominate. Most emerging countries, such as India, use more fossil fuels every year as their economies grow. It is the growing use of fossil fuel energies that has increased greenhouse gases dramatically.

Evidence – strong evidence of energy expansion in China

Explanation – ties the growth in use of fossil fuels to greenhouse gases

Point – links energy extraction as an energy consumer in itself

Not only does energy consumption increase with economic growth, but even obtaining energy is in itself responsible for increased greenhouse gas emissions. Many methods of extracting unconventional energy sources generate high levels of carbon emissions, such as tar sands, oil shale and fracking. 'Peak oil' is less of a concern now that unconventional sources have made such an impact in reducing oil process globally; the general trend of global oil prices since 2008 has been downward. US oil prices have fallen sharply since 2012, due to huge new supplies of oil and shale gas from the USA and Canada. To compete, OPEC cut its prices to maintain its market share.

Explanation – expands the point by referring to unconventional energy sources

Evidence – exemplifies oil prices in the USA and Canada

Point – strong point made about linking the burning of fossil fuels to climate change

Most assessments, for example by the IPCC, identify the burning of fossil fuels as the major cause of anthropogenic climate change and the ways in which it impacts on other systems, such as ocean temperatures. Earth's carbon reservoirs act as sources and sinks. If sources and sinks are equal, the carbon cycle is in equilibrium. However, human activities have led to increased CO_2 emissions, without corresponding increases in natural sinks (e.g. oceans and forests). Increasing atmospheric stores of carbon are widely believed to be the main cause of rising global temperatures, and fossil fuel combustion has altered the balance of these carbon pathways and stores. With carbon being released in larger amounts by burning, flows have greatly increased while stores on Earth have decreased.

Explanation – very strong explanation about impacts upon the carbon cycle

Point – switches argument to identify changes in land use as a disruption to the carbon cycle

Burning fossil fuels has important consequences for the working of the carbon cycle, but land-use changes also disrupt the way the cycle works. Land-use change, such as deforestation for palm oil or for mining, causes major changes to the chemistry and temperature of oceans, which may disrupt their role as stores in the carbon cycle. Other factors, such as peat extraction and land-use changes (e.g. forest clearance for cattle farming) are also significant. Deforestation has increased in rainforests especially, followed by the proliferation of palm oil plantations for biofuel production. Along with rainforest clearance, increased intensification of farming has also led to land-use changes, which impact on energy usage.

Evidence – uses examples of land-use changes to illustrate the point

Explanation – expands the point about land-use change further

So the statement that the most important factor in modifying the carbon cycle is the world's growing demand for energy is true, and is the main factor driving changes to the carbon cycle, though land-use changes are also significant.

Link – the candidate concludes the answer with an answer to the question

Examiner feedback

This is a strong answer, but it needs further ongoing assessment to hit the top of Level 4. The descriptor for Level 4 generally applies to all parts.

For AO1:

- The candidate's knowledge of economic growth and rising energy demands as a cause of changes to the carbon cycle is wide ranging and accurate. It meets all parts of the descriptor for Level 4, i.e. knowledge and understanding, which is detailed, thorough, relevant, accurate in explaining key concepts and processes, and has an awareness of geographical scale and change over time.

For AO2:

- The candidate's argument about land-use change as a contributor to climate change is sound, as is fossil fuels consumption. There is good use of evidence of the use of fossil fuels in China and India, and of land-use change.
- However, assessment is a little weaker. The final conclusion is brief and needs to return to the wording of the question to establish an answer. Each paragraph needs a mini-assessment to return to the question, so that the final conclusion isn't just a 'bolt-on' at the end.

By mostly meeting the Level 4 descriptors, the answer earns 17 marks in the lower part of Level 4. Had there been greater assessment, this could achieve the full 20 marks!

In this section you'll learn how to...

- maximise marks on 20-mark questions on Hot desert systems and landscapes, which use the command word 'Evaluate'.

Tackling 20-mark questions about Hot desert systems and landscapes

This topic is assessed on Paper 1, Section B, Question 2.

- In Paper 1, Section B you should answer **either** Question 2 (*Hot desert systems and landscapes, pages 71–79*) **or** Question 3 (*Coastal systems and landscapes, pages 80–88*) **or** Question 4 (*Glacial systems and landscapes, pages 89–97*).
- Before beginning this section, you should study section 1.7 carefully which will tell you about the skills you need to answer 20-mark questions, and the mark scheme.

Try this 20-mark question on Hot desert systems and landscapes

> This question assesses AO1 for 10 marks and AO2 for 10 marks.

Question

Evaluate the statement that 'water is a far more effective influence in arid landscapes than wind'. (20 marks)

1 Plan your answer

Before attempting to answer the question, remember to **BUG** it. That means:

✓ **Box** the command word.
✓ **Underline** the following:
  - the theme
  - the focus
  - any evidence required
  - the number of points needed.
✓ **Glance** back over the question – to make sure you include everything in your answer.

Use the **BUG** on the next page to plan your own answer.

Five steps to success!

Five steps to help you write top quality answers

The following five steps are used in this book to help you get the best marks.

1. **Plan your answer** – decide what to include and how to structure your answer.

2. **Write your answer** – use the answer spaces to write your answer.

3. **Mark your answer** – use the mark scheme (section 1.7) to self- or peer-mark your answer. You can also use this to assess sample answers in step 4 below.

4. **Sample answers** – two sample answers are given to show you how to maximise marks for a question.

5. **Marked sample answer** – this is the same answer that you used for step 4 above, but is marked and annotated, so that you can compare with your own answers.

Command word: 'Evaluate' means that you need to measure the extent to which the statement in the exam question is true, and provide a balanced judgement.

Evidence: You need evidence – that is, examples you have learned about – in order to help you develop an argument and make a judgement.

Evaluate the statement that 'water is a far more effective influence in arid landscapes than wind'. (20 marks)

Focus: The focus for this question is the influence of water versus wind erosion, so you need to understand the concepts involved.

Number of points: 20 marks. You need five or six reasoned points, preferably balanced across both sides of the argument (AO2), supported by evidence (AO1).

PEEL your answer

Use PEEL notes to structure your answer. This will help to explain your ideas to the examiner clearly. PEEL has four stages:

- **P**oint – Make five or six developed points for 20 marks. These are the **AO2** parts of the answer where you argue how landscapes have been modified or affected by changes, and show which points are significant.
- **E**vidence – Include details about specific ways in which hot desert landscapes have been modified by water and wind. It's important to use examples of processes and landforms produced by both.
- **E**xplain – Give a reason for each point. Use starter phrases such as: '*This particular factor is important because ...*'. This extends your **AO2** ability because you'll show how these changes can be explained.
- **L**ink back to the question – Use the wording of the question in your answer to show you recognise what the question is asking, e.g. '*This example shows the significance of water because ...*'.

 Tip

Remember, quality not quantity

You will not be marked on the **number** of points you make, but on the **quality** of your answer. That means the quality of the argument (AO2), the evidence you give (AO1), and how you link back to the question.

Remember the AOs!

It's important to know the AOs for this question.
- **AO1** (10 marks) is about demonstrating knowledge and understanding of hot desert landscape systems. It means that you need examples of processes and landforms to show the influence of water and wind in desert landscapes. Use brief examples – not long case studies.
- **AO2** (10 marks) is about using evidence to decide whether water or wind is the greater influence in desert landscapes.

2 Write your answer

Using file paper, write your answer to the following question.

Question

> Evaluate the statement that 'water is a far more effective influence in arid landscapes than wind'. (20 marks)

3 Mark your answer

1. To help you to identify well-structured points, highlight or underline:

- points in red
- explanations in orange
- evidence in blue
- links back to the question by underlining.

Look for an introduction and conclusion, which are essential for questions with the command word 'Evaluate'.

2. Use the mark scheme in section 1.7. Remember that:

- 20-mark questions are marked by choosing a level and a mark within it, based upon the answer as a whole
- a high-scoring answer must include **both** AOs. One which includes only AO1 without applying the question for AO2 can only gain a maximum of 10 marks.

Things to watch out for

Study Sample Answer 1 on the next page.

- The candidate has done some things well, so identify which qualities are good.
- Identify what prevents the answer from getting a higher mark.

  Clues:
  Look at how the candidate demonstrates:
  a) **AO1** – what does the candidate know about water and wind processes in hot desert landscapes?
  b) **AO2** – how well does the candidate develop arguments about which of the two influences is more effective and why? Are the two sets of processes assessed for their effectiveness, and compared?

Remember the introduction and conclusion!

All questions with the command word 'Evaluate' need to have:
- a brief (1–2 sentence) introduction
- a final short paragraph conclusion.

Sample Answer 1

Hot deserts are mostly found in tropical areas and in continental interiors, like the Sahara or Gobi deserts. They vary as some are sandy like the northern Sahara, while others are rocky like the deserts of Arizona. The processes that form these desert landscapes include wind and water, which are also important in transporting sediment. I will compare different features of hot deserts to try and show whether wind or water is more effective.

Most hot desert areas have really high temperatures during the day with 50 °C or more not uncommon. Skies are clear because the air is so dry, and so temperatures fall rapidly at night and therefore frost is not uncommon. This means that weathering of rocks in hot desert climates puts stresses on rocks which expand in the high temperatures and contract at night. This weakens them and causes exfoliation of layers of rock parallel to the surface. Sometimes if frost occurs then any water in the rock freezes and causes freeze-thaw which prises away pieces of rock to form scree. These processes are examples of weathering.

Wind is very important in forming desert landforms. Some are formed when wind picks up sediment and transports it away – these are called deflation hollows. They sometimes leave bare rock with no sand left. Sand gets blown away and then gets deposited in other places. Where it gets deposited is where sand dunes form. Sand dunes form as wind deposits and then get shaped by wind – the sand accumulates to form high ridges and then, where the wind blows around the side, it tapers. These are called crescent dunes or barchans.

Rock deserts also form in places where there is no soil or surface covering. Bare rock surfaces can be flat, desert pavements. But there are sometimes rock features called yardangs and zeugen. These are quite similar and consist of a large rock feature sitting above the rock surface, but with a narrow pillar attaching it to the ground surface. This is narrow because it has probably been eroded by the wind which blasts away and causes abrasion at the base. This creates the pillar on which the rock stands.

Water is also important in forming hot desert landscapes, even though it rains infrequently. The rain can form in sudden storms which cause surface runoff because the rain is falling faster than the ground is able to absorb it. This means that there can be very sudden flash floods and very sudden dangerous rivers can form in steep sided river valleys, known as wadis. These are deep because water is able to erode them during these storms. Flooding in deserts is more common than you would think.

So in conclusion wind and water are both important in hot desert landscapes. Wind is probably there more of the time but water has a big impact when storms occur.

	Remember! Check for both AO1 (knowledge and understanding) and AO2 (application).	
Strengths of the answer		
Ways to improve the answer		
Level		**Mark**	

Examiner feedback

As with all 20-mark questions, there are four levels for this question. The style of the answer is a little lacking in detail, but the candidate has created an answer in which there are some good, relevant points.

- The candidate has some fairly good knowledge and understanding (AO1) of landforms and the climate of hot desert regions. Examples are given.
- The paragraphs are organised between weathering processes, and wind and water landforms. There is some knowledge of the landscape processes involved.
- To earn a higher mark, the candidate needs to develop more application to the question. There is a brief conclusion at the end but it needs to be built up more throughout the answer – it is rather bolted on. The candidate needs to try to include brief 'mini-conclusions' at the end of each paragraph before a main conclusion at the end.

For AO1, the descriptor for Level 3 applies:

- *Knowledge and understanding is mostly clear, accurate and relevant to the argument, in explaining concepts and processes, and in explaining geographical scale and change over time.*

For AO2, Level 3 of the mark scheme applies to the first part:

- *Application is sound in different contexts (e.g. in different examples used), shows thorough, clear, and relevant analysis,*

However, the conclusion is weak, so Level 2 is more appropriate:

- *Leads to a generally sound, evaluative conclusion (though with little ongoing evaluation throughout the answer) which is sometimes based on evidence.*

The answer therefore fits the criteria for Level 3 overall, though not fully, because the conclusion is not well developed. Therefore an examiner would award this low Level 3 for 12 marks.

4 # Now mark this one!

Read through Sample Answer 2 on the next page.

a) Go through the answer using the three colours in section 3, and underline any links back to the question to assess the effectiveness of water and wind.
b) Look for evidence of both knowledge and application.
c) Look for an introduction and conclusion, which are essential for questions with the command word 'Evaluate'.
d) Use the mark scheme in section 1.7 to decide which level it is in and how many marks it is worth.

 Question recap

Question 2 Evaluate the statement that 'water is a far more effective influence in arid landscapes than wind'. (20 marks)

When people think of hot deserts, images that come to mind are often those of sand landscapes such as dunes, or perhaps dry rocky surfaces like those shown in American films, with rock features in the far distance. These might imply that wind is the most effective influence in hot desert landscapes, because sand grains are easily moved by moving air, and dust storms occur frequently. But many desert landforms actually require water in their formation and development. In this answer I shall compare several landscape features of hot deserts in order to evaluate whether wind or water is more effective.

The source of energy in hot desert environments is the sun's insolation, creating convection currents of air and winds, some of which bring occasional moisture. For most hot deserts, aridity is the key climatic characteristic with infrequent rainfall, low annual totals of precipitation and few days when rain falls. Plant characteristics have evolved around either drought avoidance or drought resistance. All these factors would seem to point to wind and drought as ever-present, compared to moisture availability.

Both erosion and deposition by wind are important in creating desert landscapes. Landforms associated with wind have several unique characteristics. Deflation hollows are created by wind deflation, transporting sand away and creating a net removal of sediment. Deposition creates different forms of sand dune, depending upon the sediment budget available and wind patterns. Seif dunes have evolved in the Namib desert where local winds 'tunnel' their way between ridges of sand, developing long, thin dunes during one season, whilst a seasonal change of wind direction results in winds blowing at right angles to these dunes for the rest of the year. These contrast from the typical crescent-shaped barchan dunes, formed with the direction of a single dominant prevailing wind.

As well as sandy deserts, rock deserts are also common, such as desert pavements. Like deflation hollows, these have been stripped of sand, usually by wind, and are common in the Sahara. There is no doubt that events such as sandstorms are capable of transporting vast amount of sediment, often to great heights in the atmosphere.

However, other features such as yardangs and zeugen are more problematic. These features were commonly thought at one time to illustrate the effectiveness of wind in shaping desert landforms. Weaker rock strata in yardangs and zeugen, for example, were held to be the result of sustained abrasion by wind of those strata. However, even early research over a century ago identified that weaker strata were likely to be weakened by chemical weathering, and so even with wind abrasion, water is effective in the development of these landforms. Often wind-borne particles most capable of abrasion are only carried to shallow heights above ground level, making it unlikely that wind plays a major part in the formation of these landforms.

While all of these features are significant in identifying the effectiveness of wind in developing hot desert landscape features, the role of water cannot be understated. Water may be present during brief, episodic events but its impact is immense. Surface runoff in sudden storms is often rapid or even immediate, so that many roads across Australia's deserts warn of flash and sheet floods in what appears to be an arid landscape. Water may be present only infrequently, but drainage patterns can clearly be seen from the air across deserts like the Kalahari, with networks of streams marked out by increased densities of vegetation. Many such river networks never reach the sea but are internal drainage basins. Sudden flash floods contained within desert gorges, or wadis, are extremely effective in erosion because the lack of vegetation allows them to pick up sediment and, though the river may be short-lived (as water both infiltrates and evaporates), it carries huge erosive energy in a short time. Similarly, deposition occurs, especially where wadis end and rivers choked with sediment fan out across broader plains, forming bahadas.

While wind may play a significant role in the development of desert landforms, there is no doubt that water plays at least as important a role, if not more so. Many landforms owe their origin to water in the landscape.

	Remember! Check for both AO1 (knowledge and understanding) and AO2 (application).
Strengths of the answer	
Ways to improve the answer	

Level		**Mark**	

5 Marked sample answers

Sample Answer 2 is marked on the next page. The text has been highlighted to show how well each answer has structured points.

The following have been highlighted:

- points in red
- explanations in orange
- evidence in blue
- underlined points are where the candidate links to the question to evaluate the importance of wind or water.

Marked sample answer 2

Link – the candidate introduces the answer well, defining the issue within the question, but not the key terms

When people think of hot deserts, images that come to mind are often those of sand landscapes such as dunes, or perhaps dry rocky surfaces like those shown in American films, with rock features in the far distance. These might imply that wind is the most effective influence in hot desert landscapes, because sand grains are easily moved by moving air, and dust storms occur frequently. But many desert landforms actually require water in their formation and development. In this answer I shall compare several landscape features of hot deserts in order to evaluate whether wind or water is more effective.

Point – refers first to the sun as the source of energy driving the whole system

The source of energy in hot desert environments is the sun's insolation, creating convection currents of air and winds, some of which bring occasional moisture. For most hot deserts, aridity is the key climatic characteristic with infrequent rainfall, low annual totals of precipitation and few days when rain falls. Plant characteristics have evolved around either drought avoidance or drought resistance. All these factors would seem to point to wind and drought as ever-present, compared to moisture availability.

Explanation – extends the explanation and outlines impacts of changes

Evidence – uses evidence to illustrate the characteristics of desert climates

Link – links back to the question to establish the influence of wind

Both erosion and deposition by wind are important in creating desert landscapes. Landforms associated with wind have several unique characteristics. Deflation hollows are created by wind deflation, transporting sand away and creating a net removal of sediment. Deposition creates different forms of sand dune, depending upon the sediment budget available and wind patterns. Seif dunes have evolved in the Namib desert where local winds 'tunnel' their way between ridges of sand, developing long, thin dunes during one season, whilst a seasonal change of wind direction results in winds blowing at right angles to these dunes for the rest of the year. These contrast from the typical crescent-shaped barchan dunes, formed with the direction of a single dominant prevailing wind.

Explanation – extends this using processes created by wind

Point – the candidate switches to rock deserts

Point – introduces characteristics of wind features

Evidence – exemplifies landform features created by wind

As well as sandy deserts, rock deserts are also common, such as desert pavements. Like deflation hollows, these have been stripped of sand, usually by wind, and are common in the Sahara. There is no doubt that events such as sandstorms are capable of transporting vast amount of sediment, often to great heights in the atmosphere.

Explanation – extends the explanation to illustrate the impact of wind

Evidence – evidence of wind process is given to illustrate the point

Point – switches attention to landforms that pose questions about wind

Explanation – compares the impacts of wind processes to suggest these are the unlikely cause

Evidence – substantial evidence given to illustrate the point

Link – provides a brief conclusion answering the question

However, other features such as yardangs and zeugen are more problematic. These features were commonly thought at one time to illustrate the effectiveness of wind in shaping desert landforms. Weaker rock strata in yardangs and zeugen, for example, were held to be the result of sustained abrasion by wind of those strata. However, even early research over a century ago identified that weaker strata were likely to be weakened by chemical weathering, and so even with wind abrasion, water is effective in the development of these landforms. Often wind-borne particles most capable of abrasion are only carried to shallow heights above ground level, making it unlikely that wind plays a major part in the formation of these landforms.

While all of these features are significant in identifying the effectiveness of wind in developing hot desert landscape features, the role of water cannot be understated. Water may be present during brief, episodic events but its impact is immense. Surface runoff in sudden storms is often rapid or even immediate, so that many roads across Australia's deserts warn of flash and sheet floods in what appears to be an arid landscape. Water may be present only infrequently, but drainage patterns can clearly be seen from the air across deserts like the Kalahari, with networks of streams marked out by increased densities of vegetation. Many such river networks never reach the sea but are internal drainage basins. Sudden flash floods contained within desert gorges, or wadis, are extremely effective in erosion because the lack of vegetation allows them to pick up sediment and, though the river may be short-lived (as water both infiltrates and evaporates), it carries huge erosive energy in a short time. Similarly, deposition occurs, especially where wadis end and rivers choked with sediment fan out across broader plains, forming bahadas.

While wind may play a significant role in the development of desert landforms, there is no doubt that water plays at least as important a role, if not more so. Many landforms owe their origin to water in the landscape.

Evidence – gives exemplars of how processes may be different from how they seem

Point – established water as a powerful influence

Explanation – extended explanations about impacts of water in the landscape

Examiner feedback

This is a strong answer. The descriptor for Level 4 applies to all parts except one:

For AO1:

- The candidate's knowledge of desert landscape features and examples is wide ranging and accurate. It meets all parts of the descriptor for Level 4 – i.e. *knowledge and understanding* which is detailed, thorough, relevant to the argument, accurate in explaining key concepts and processes, and has an awareness of geographical scale and change over time.

AO2 has all the qualities for Level 4 except one.

- *Application* is thorough in a range of examples given about wind and water-formed features, and the *analysis* is detailed, coherent and relevant.
- However, evaluation is weaker. The final conclusion is brief and needs to return to the wording of the question to establish an answer. Only 2 or 3 sentences are needed. There is just one link showing evidence of ongoing evaluation, and there needs to be more like this.

By mostly meeting the Level 4 descriptors, the answer earns 17 marks in the lower part of Level 4. Had there been an evaluative conclusion, this could achieve the full 20 marks!

- maximise marks on 20-mark questions on Coastal systems and landscapes, which use the command word 'Evaluate'.

Tackling 20-mark questions about Coastal systems and landscapes

This topic is assessed on Paper 1, Section B, Question 3.

- In Paper 1, Section B you should answer **either** Question 2 (*Hot desert systems and landscapes, pages 71–79*) **or** Question 3 (*Coastal systems and landscapes, pages 80–88*) **or** Question 4 (*Glacial systems and landscapes, pages 89–97*).
- Before beginning this section, you should study section 1.7 carefully which will tell you about the skills you need to answer 20-mark questions, and the mark scheme.

Try this 20-mark question on Coastal systems and landscapes

> This question assesses AO1 for 10 marks and AO2 for 10 marks.

Question

Evaluate the ways in which present-day coastal landscapes have been modified or affected by changes in sea level. (20 marks)

1 Plan your answer

Before attempting to answer the question, remember to **BUG** it. That means:

✓ **Box** the command word.
✓ **Underline** the following:
  - the theme
  - the focus
  - any evidence required
  - the number of points needed.
✓ **Glance** back over the question – to make sure you include everything in your answer.

Use the **BUG** on the next page to plan your own answer.

Five steps to success!

Five steps to help you write top quality answers

The following five steps are used in this book to help you get the best marks.

1. **Plan your answer** – decide what to include and how to structure your answer.

2. **Write your answer** – use the answer spaces to write your answer.

3. **Mark your answer** – use the mark scheme (section 1.7) to self- or peer-mark your answer. You can also use this to assess sample answers in step 4 below.

4. **Sample answers** – two sample answers are given to show you how to maximise marks for a question.

5. **Marked sample answer** – this is the same answer that you used for step 4 above, but is marked and annotated, so that you can compare with your own answers.

Command word: 'Evaluate' means that you need to measure the extent to which the statement in the exam question is true, and provide a balanced judgement.

Evidence: You need evidence – that is, examples you have learned about – in order to help you develop an argument and make a judgement.

Evaluate the ways in which present-day coastal landscapes have been modified or affected by changes in sea level. (20 marks)

Focus: The focus for this question is 'changes in sea level', so you need to understand the concepts involved.

Number of points: 20 marks. You need five or six reasoned points, preferably balanced across both sides of the argument (AO2), supported by evidence (AO1).

PEEL your answer

Use PEEL notes to structure your answer. This will help to explain your ideas to the examiner clearly. PEEL has four stages:

- **P**oint – Make five or six developed points for 20 marks. These are the **AO2** parts of the answer where you argue how landscapes have been modified or affected by changes, and show which points are significant.
- **E**vidence – Include details about specific ways in which coastal landscapes have been modified by change to illustrate your argument. It's important to use examples, such as the *coastal landscapes affected by sea level change.*
- **E**xplain – Give a reason for each point. Use starter phrases such as: '*This particular factor is important because ...*'. This extends your **AO2** ability because you'll show how these changes can be explained.
- **L**ink back to the question – Use the wording of the question in your answer to show you recognise what the question is asking, e.g. '*This example shows the significance of changes in sea level because ...*'.

 Tip

Remember, quality not quantity

You will not be marked on the **number** of points you make, but on the **quality** of your answer. That means the quality of the argument (AO2), the evidence you give (AO1), and how you link back to the question.

Remember the AOs!

It's important to know the AOs for this question.

- **AO1** (10 marks) is about demonstrating knowledge and understanding of coastal landscape systems. It means that you need examples of changes in sea level. Use brief examples – not long case studies.
- **AO2** (10 marks) is about using evidence to decide which sea level changes have been most significant in altering coastal landscapes.

2 Write your answer

Using file paper, write your answer to the following question.

Question

Evaluate the ways in which present-day coastal landscapes have been modified or affected by changes in sea level. (20 marks)

3 Mark your answer

1. To help you to identify well-structured points, highlight or underline:

- points in red
- explanations in orange
- evidence in blue
- links back to the question by underlining.

Look for an introduction and conclusion, which are essential for questions with the command word 'Evaluate'.

2. Use the mark scheme in section 1.7. Remember that:

- 20-mark questions are marked by choosing a level and a mark within it, based upon the answer as a whole
- a high-scoring answer must include **both** AOs. One which includes only AO1 without applying the question for AO2 can only gain a maximum of 10 marks.

Things to watch out for

Study Sample Answer 1 below.

- The candidate has done some things well, so identify which qualities are good.
- Identify what prevents the answer from getting a higher mark.

  Clues:
  Look at how the candidate demonstrates:
  a) **AO1** – what does the candidate know about sea level change and ways in which it can alter coastal landscapes?
  b) **AO2** – how well does the candidate develop arguments about which specific present-day coastal landscapes have been modified or affected by changes in sea level? Are particular changes in sea level assessed for their impact on coastal landscapes, compared to others?

Remember the introduction and conclusion!

All questions with the command word 'Evaluate' need to have:
- a brief (1–2 sentence) introduction
- a final short paragraph conclusion.

Sample Answer 1

Present-day coastal landscapes have been modified or affected by changes in sea level. Some of these processes are long term and it is necessary to go back to the Ice Age for causes of changes. After the Ice Age 10 000 years ago, glaciers and ice caps melted as the climate got warmer. This caused a large rise in sea level throughout the world which caused coastal submergence. There are examples of submerged coasts in the UK like fjords and lochs near the sea in Scotland and in Norway. These are called eustatic changes caused by the sea.

Eustatic changes can be slow or rapid. At the moment, sea level seems to rise by about 3.2mm every year. This is very rapid compared to historical processes in the past and is mostly due to climate change caused by anthropogenic factors such as burning fossil fuels, which has increased the concentrations of greenhouse gases in the atmosphere. This has led to greater threats of coastal flooding in countries such as Bangladesh, where 40% of the country lies at 1 metre or less above sea level. Increased coastal flooding and inundation would alter the landscape and would be a massive threat to the population. Lowland parts of the UK and Europe are also under threat so the landscape would also change there and make urban areas such as London under threat from inundation.

But there have been times when sea level has fallen or land has risen above the sea. The glaciers and ice sheets were heavy and weighed down on the crust, so it recovered by rising sharply above sea level when the glaciers melted and there was less weight. This is called isostatic change. There are landforms like raised beaches in Scotland which have been lifted above the sea – they used to be at sea level but isostatic change lifted them higher. These consist of beach materials which are now several metres above the sea, with features such as shoreline platforms raised above the current coast.

Many landscapes consist of landforms that are developing at the present time. Beaches change almost daily as tides bring swash and backwash over them. They also change due to longshore drift, which is different at high tide and low tide, so a beach changes in that way too.

Storms can cause major short-term change in sea level because low pressure systems cause sea level to rise, so that high tides are higher than usual and can cause coastal flooding. They can also increase erosion rates on cliffs and beaches when sea level becomes so much higher. Some of the storms of the winter of 2013–14 caused changes to the coastline, as in Dawlish in Devon where a sand replenishment scheme from the previous summer was washed away by one storm. This shows how coasts can be modified overnight just by tidal changes in a big storm.

So changes to landscapes on the coast can result from climate changes which cause an increase or decrease in sea level.

	Remember! Check for both AO1 (knowledge and understanding) and AO2 (application).	
Strengths of the answer		
Ways to improve the answer		
Level		**Mark**	

Examiner feedback

As with all 20-mark questions, there are four levels for this question. The candidate has created an answer in which there are some relevant points.

- The candidate has some knowledge and understanding (AO1) of the causes of eustatic change, and of sea level change generally. There is sound knowledge of places and events, of examples of sea level change (e.g. raised beaches), and of the causes and impacts of storms.
- There is some irrelevance – the answer contains too much on causes of climate change, and of impacts on countries like Bangladesh, and not enough on coastal landscapes resulting from sea level change.
- To earn a higher mark, the candidate needs to develop more application to the question. Few processes, such as eustatic change, are linked to coastal landscapes. Particular features are described (e.g. raised beaches), but there is no sense of a coastal landscape consisting of several features, using these to develop an answer.

For AO1, the descriptor for Level 3 applies:

- Knowledge and understanding is mostly clear, accurate and relevant to the argument, in explaining concepts and processes, and in explaining geographical scale and change over time.

For AO2, Level 3 of the mark scheme applies to the first part:

- *Application is sound in different contexts (e.g. in different examples used), shows thorough, clear, and relevant analysis,*

However, the conclusion is not well developed, so Level 2 is more appropriate:

- *leads to a generally sound, evaluative conclusion (though with little ongoing evaluation throughout the answer) which is sometimes based on evidence.*

The answer therefore fits the criteria for Level 3 overall, though not fully, because the conclusion is not well developed. Therefore an examiner would award this low-middle Level 3 for 13 marks.

④ Now mark this one!

Read through Sample Answer 2 below.

    a) Go through the answer using the three colours in section 3, and underline any links back to the question.
    b) Look for evidence of both knowledge and understanding.
    c) Use the mark scheme in section 1.7 to decide which level it is in and how many marks it is worth.

 Question recap

Question 3 Evaluate the ways in which present-day coastal landscapes have been modified or affected by changes in sea level. (20 marks)

Sample Answer 1

Most present-day coastal landscapes result from a combination of factors. Classic landforms such as arches may depend on the geology of a coastline, as well as energy levels of waves. But many coastlines depend on sea level changes, some long-term and some short-term, and this answer will try to evaluate the contribution that these changes make to coastal landscapes.

After the last Ice Age, about 10 000 years ago, the melting of both glaciers and ice caps caused widespread submergence of coasts as sea levels rose, known as eustatic changes. Examples of submerged coasts in the UK include the ria coastlines of Devon and Cornwall, where deep water channels now occupy what had been steep-sided river valleys during the Ice Age. The flooding of these valleys changed their shape and form, so that the incursions of sea water in the Dart estuary in south Devon resemble the dendritic patterns of tributaries in river valleys.

This rise in sea level was relative, and in some areas the reduced weight of ice sheets and glaciers on continental crust caused a rise in land levels known as isostatic readjustment. In some parts of the UK, e.g. the west coast of Scotland, isostatic changes exceeded those caused by rising sea levels. North of Oban lies a coastal stack in a field, about 10 metres above sea level, and located inland from the present position of the coastline. Its isolation from the coast results from isostatic change, which has exceeded the rise of eustatic change. However, not all of Scotland has adjusted in this way because sea lochs have flooded in much the same way as rias, like Norwegian fjords.

Eustatic and isostatic changes are fairly long term. Many coastal landscapes also consist of features that have developed over the shorter term. These vary between beach cusps, which can result from daily tidal changes or wind changes along a beach over a few days. Other beach features result from one-off sudden events, such as storms, so that a berm may grow or be relocated depending on the force of a storm. The nature of short-term changes depends upon the topography of the coastline. Single storms, or those that are repeated over winter months in the UK, can result in spectacular cliff collapses like those which occurred along the chalk coast of Kent near Dover in winter 2014. Similarly, coastal bars can be breached as a result of a single storm, such as that at Tor Sands near Slapton in south Devon in winter 2018.

Similarly, storm surges, caused by short-term reductions in air pressure, increase sea levels and can have a significant impact on the creation of landforms, which can be dramatic, such as coastal inundation which occurred during Hurricane Katrina. Tectonic activity can have similar impacts, like the submersion of parts of Banda Aceh in Indonesia which resulted from the earthquake which led to the 2004 Asian tsunami.

Other changes occur over a longer term, such as the formation of headland and bays. Here, changes in sea level have helped to determine coastal landscapes in the formation of bays. The disposition of strata, with alternate weak and resistant rock types, affects this process. But past processes also play their part. Lulworth Cove in Dorset is widely believed to have been eroded and formed by the local river valley during the periglacial period as well as wave breaches of the marine cliffs.

In conclusion, it is clear that present-day coastal landscapes have been modified or affected significantly by changes in sea level. However, a time dimension is needed to qualify this statement, since those changes that result from short-term changes over a single day, month or season are often both temporary and insignificant, compared to longer-term changes brought by sustained changes to sea level.

	Remember! Check for both AO1 (knowledge and understanding) and AO2 (application).	
Strengths of the answer		
Ways to improve the answer		
Level		**Mark**	

⑤ Marked sample answers

Sample Answer 2 is marked below. The text has been highlighted to show how well each answer has structured points.

The following have been highlighted:

- points in red
- explanations in orange
- evidence in blue
- underlined points are where the candidate links to the question to identify the importance of particular changes in sea level.

Look for the introduction and conclusion, which are essential for questions with the command word 'Evaluate'.

Marked sample answer 2

> Link – the candidate introduces the answer well, using the wording of the question, and defines both key terms, as well as setting out the argument

Most present-day coastal landscapes result from a combination of factors. Classic landforms such as arches may depend on the geology of a coastline, as well as energy levels of waves. But many coastlines depend on sea level changes, some long-term and some short-term, and this answer will try to evaluate the contribution that these changes make to coastal landscapes.

After the last Ice Age, about 10 000 years ago, the melting of both glaciers and ice caps caused widespread submergence of coasts as sea levels rose, known as eustatic changes. Examples of submerged coasts in the UK include the ria coastlines of Devon and Cornwall, where deep water channels now occupy what had been steep-sided river valleys during the Ice Age. The flooding of these valleys changed their shape and form, so that the incursions of sea water in the Dart estuary in south Devon resemble the dendritic patterns of tributaries in river valleys.

This rise in sea level was relative, and in some areas the reduced weight of ice sheets and glaciers on continental crust caused a rise in land levels known as isostatic readjustment. In some parts of the UK, e.g. the west coast of Scotland, isostatic changes exceeded those caused by rising sea levels. North of Oban lies a coastal stack in a field, about 10 metres above sea level, and located inland from the present position of the coastline. Its isolation from the coast results from isostatic change, which has exceeded the rise of eustatic change. However, not all of Scotland has adjusted in this way because sea lochs have flooded in much the same way as rias, like Norwegian fjords.

Evidence – gives examples of submerged coasts in the UK

Point – the candidate balances submergence with isostatic readjustment

Evidence – gives an example of a landform formed by isostatic change

Point – the candidate refers first to submergence of coastlines

Explanation – the candidate explains the impact of submergence in creating rias

Explanation – the candidate explains how this affected coastal landscapes

Explanation – explains the impact of storms on coastlines

Evidence – exemplars are given about the impact of storms on stretches of coast and landforms

Point – that sea level changes can occur as a result of storms and tectonic events

Evidence – two exemplars are given

Point – the candidate makes the point about the influence of geology and how sea level changes affect headlands and bays

Evidence – uses the example of Lulworth Cove to explain this

Point – the candidate switches from long-term to short-term changes

Explanation – these impacts are explained

Explanation – the candidate extends this further but evaluates this against the contribution of river processes in the formation of bays

Eustatic and isostatic changes are fairly long term. Many coastal landscapes also consist of features that have developed over the shorter term. These vary between beach cusps, which can result from daily tidal changes or wind changes along a beach over a few days. Other beach features result from one-off sudden events, such as storms, so that a berm may grow or be relocated depending on the force of a storm. The nature of short-term changes depends upon the topography of the coastline. Single storms, or those that are repeated over winter months in the UK, can result in spectacular cliff collapses like those which occurred along the chalk coast of Kent near Dover in winter 2014. Similarly, coastal bars can be breached as a result of a single storm, such as that at Tor Sands near Slapton in south Devon in winter 2018.

Similarly, storm surges, caused by short-term reductions in air pressure, increase sea levels and can have a significant impact on the creation of landforms, which can be dramatic, such as coastal inundation which occurred during Hurricane Katrina. Tectonic activity can have similar impacts, like the submersion of parts of Banda Aceh in Indonesia which resulted from the earthquake which led to the 2004 Asian tsunami.

Other changes occur over a longer term, such as the formation of headland and bays. Here, changes in sea level have helped to determine coastal landscapes in the formation of bays. The disposition of strata, with alternate weak and resistant rock types, affects this process. But past processes also play their part. Lulworth Cove in Dorset is widely believed to have been eroded and formed by the local river valley during the periglacial period as well as wave breaches of the marine cliffs.

In conclusion, it is clear that present-day coastal landscapes have been modified or affected significantly by changes in sea level. However, a time dimension is needed to qualify this statement, since those changes that result from short-term changes over a single day, month or season are often both temporary and insignificant, compared to longer-term changes brought by sustained changes to sea level.

Link – provides a sound conclusion referring to the wording of the question

Examiner feedback

The descriptor for Level 4 applies to all parts:

For AO1:

- The candidate's knowledge and explanations of sea level changes and their impacts upon coastlines (together with examples) is wide ranging and accurate. It meets all parts of the descriptor for Level 4 – i.e. *knowledge and understanding* which is detailed, thorough, relevant to the argument, accurate in explaining key concepts and processes, and has an awareness of geographical scale and change over time.

AO2 has all the qualities for Level 4 except one. Application is thorough in a range of examples given about coastal landscapes resulting from sea level change, and the analysis is detailed, coherent and relevant. There is a logical conclusion based on evidence.

By mostly meeting the Level 4 descriptors, the answer earns 18 marks in the middle of Level 4. To reach full marks, the candidate needs to have:

- discussed more what coastal *landscapes* look like – most of the exemplars are *landforms*, so a slightly broader view is needed. The section which explains landscapes well is the part of the answer about rias, because these are described as a whole.

In this section you'll learn how to...

- maximise marks on 20-mark questions on Glacial systems and landscapes, which use the command word 'Evaluate'.

Tackling 20-mark questions about Glacial systems and landscapes

This topic is assessed on Paper 1, Section B, Question 4.

- In Paper 1, Section B you should answer **either** Question 2 (*Hot desert systems and landscapes, pages 71–79*) **or** Question 3 (*Coastal systems and landscapes, pages 80–88*) **or** Question 4 (*Glacial systems and landscapes, pages 89–97*).
- Before beginning this section, you should study section 1.7 carefully which will tell you about the skills you need to answer 20-mark questions, and the mark scheme.

Try this 20-mark question on Glacial systems and landscapes

> This question assesses AO1 for 10 marks and AO2 for 10 marks.

Question

Evaluate the statement that 'while glacial landscapes show ample evidence of global warming, it is actually periglacial landscapes that are most vulnerable to changes in climate'. (20 marks)

1 Plan your answer

Before attempting to answer the question, remember to **BUG** it. That means:

✓ **Box** the command word.
✓ **Underline** the following:
  - the theme
  - the focus
  - any evidence required
  - the number of points needed.
✓ **Glance** back over the question – to make sure you include everything in your answer.

Use the **BUG** on the next page to plan your own answer.

Five steps to success!

Five steps to help you write top quality answers

The following five steps are used in this book to help you get the best marks.

1. **Plan your answer** – decide what to include and how to structure your answer.

2. **Write your answer** – use the answer spaces to write your answer.

3. **Mark your answer** – use the mark scheme (section 1.7) to self- or peer-mark your answer. You can also use this to assess sample answers in step 4 below.

4. **Sample answers** – two sample answers are given to show you how to maximise marks for a question.

5. **Marked sample answer** – this is the same answer that you used for step 4 above, but is marked and annotated, so that you can compare with your own answers.

Command word: 'Evaluate' means that you need to measure the extent to which the statement in the exam question is true, and provide a balanced judgement.

Focus: The focus for this question is threats to 'periglacial landscapes', so you need to understand the processes involved.

Evaluate the statement that 'while glacial landscapes show ample evidence of global warming, it is actually periglacial landscapes that are most vulnerable to changes in climate'. (20 marks)

Evidence: You need evidence – that is, examples you have learned about – in order to help you develop an argument and make a judgement.

Number of points: 20 marks. You need five or six reasoned points, preferably balanced across both sides of the argument (AO2), supported by evidence (AO1).

PEEL your answer

Use PEEL notes to structure your answer. This will help to explain your ideas to the examiner clearly. PEEL has four stages:

- **P**oint – Make five or six developed points for 20 marks. These are the **AO2** parts of the answer where you argue how glacial landscapes have been modified and how periglacial landscapes are vulnerable.
- **E**vidence – Include details about specific ways in which glacial landscapes show evidence of climate change, as well as particular examples of the *vulnerability of periglacial landscapes*. This is **AO1** – using knowledge to support your argument.
- **E**xplain – Give a reason for each point. Use starter phrases such as: '*This particular factor is important because ...*'. This extends your **AO2** ability because you'll show how these changes can be explained.
- **L**ink back to the question – Use the wording of the question in your answer to show you recognise what the question is asking, e.g. '*This example shows the vulnerability of periglacial landscapes because ...*'.

Remember, quality not quantity

You will not be marked on the **number** of points you make, but on the **quality** of your answer. That means the quality of the argument (AO2), the evidence you give (AO1), and how you link back to the question.

Remember the AOs!

It's important to know the AOs for this question.
- **AO1** (10 marks) is about demonstrating knowledge and understanding of glacial landscape systems. It means that you need examples of glacial and periglacial landscapes. Use brief examples – not long case studies.
- **AO2** (10 marks) is about using evidence to decide whether glacial or periglacial landscapes are more vulnerable to climate change.

2 Write your answer

Using file paper, write your answer to the following question.

Question

Evaluate the statement that 'while glacial landscapes show ample evidence of global warming, it is actually periglacial landscapes that are most vulnerable to changes in climate'. (20 marks)

3 Mark your answer

1. To help you to identify well-structured points, highlight or underline:

- points in red
- explanations in orange
- evidence in blue
- links back to the question by underlining.

Look for an introduction and conclusion, which are essential for questions with the command word 'Evaluate'.

2. Use the mark scheme in section 1.7. Remember that:

- 20-mark questions are marked by choosing a level and a mark within it, based upon the answer as a whole
- a high-scoring answer must include **both** AOs. One which includes only AO1 without applying the question for AO2 can only gain a maximum of 10 marks.

Things to watch out for

Study Sample Answer 1 on the next page.

- The candidate has done some things well, so identify which qualities are good.
- Identify what prevents the answer from getting a higher mark.

    Clues:
    Look at how the candidate demonstrates:
    a) **AO1** – what does the candidate know about climate change and vulnerability in glacial and periglacial landscapes?
    b) **AO2** – how well does the candidate develop arguments about which specific changes are greatest? Are particular changes to glacial and periglacial areas assessed for their vulnerability, and compared?

Remember the introduction and conclusion!

All questions with the command word 'Evaluate' need to have:
- a brief (1–2 sentence) introduction
- a final short paragraph conclusion.

Sample Answer 1

Glacial landscapes vary hugely. A landscape is an assemblage of landforms and these landforms vary. Many factors affect the formation of these landscapes – one physical influence is geomorphological, the study of physical processes, which includes weathering and erosion, to influence the formation of physical features. Other influences include climate and atmospheric processes. I will investigate what these influences are and how glacial landscapes can be both active and relict. Glacial environments are those dominated by ice and periglacial are those influenced by ice, but on the margins of it.

Antarctica is a unique landscape that is influenced by glacial processes of erosion. These form U-shaped valleys, truncated spurs, pyramidal peaks, and arêtes. U-shaped valleys have a distinct trough shape, and are steep valleys formed by glaciers. Truncated spurs are rounded mounds of land formed by river erosion and rounded off by glaciers. Arêtes are sharp ridges formed by erosion working on different sides of a ridge, and pyramidal peaks are erected by the same erosion but where three or more sides are eroded.

Alberta in Canada is an example of a landscape influenced heavily by periglacial processes in a combination which creates a unique landscape. The influence of ice is underground rather than on the surface with glacial processes. Frost heave is an upward swelling of the ice which heaves soil upwards and creates pingos – landforms which are present in the Mackenzie River Delta. There are also solifluction lobes formed on the sides of gently sloping hills where a lack of grip of saturated soil over the permafrost causes it to slide.

These differences are all due to geomorphological processes, but climate change is also an influence. This is highly influential in the formation of landforms and a variety of landscapes arises. Milankovitch's theory suggests that changes in the Earth's orbit have caused climate change, so that relict landscapes have formed in the past. So there are landscapes formed anything up to 100 000 years ago, including the cirques and troughs found in the Lake District, and in features such as the knock and lochan landscapes of the Scottish Cairngorms. These past processes have also produced depositional landforms such as drumlins, which can be seen in Ribblesdale in the Yorkshire Dales. These were caused by changes in climate 100 000 years ago, and gradual warming which caused the ice to melt probably about 10 000 years ago. So the statement is true about global warming because the Ice Age has now disappeared and the earth is in a warm phase.

The view that periglacial landscapes are most vulnerable to changes in climate is true because periglacial features formed in the last Ice Age and can now only be seen in areas like the Arctic. Now the Arctic is warming rapidly because of climate change and the tundra is changing. Trees are growing now where only small shrubs and lichens could grow because of permafrost. In summer a lot of the tundra is saturated, so solifluction and other geomorphological processes occur more rapidly, which is changing the landscape. But glacial areas are changing too because glaciers are melting more than they are accumulating. So the statement that 'while glacial landscapes show ample evidence of global warming, it is actually periglacial landscapes that are most vulnerable to changes in climate' is partly true but glacial areas are vulnerable too.

	Remember! Check for both AO1 (knowledge and understanding) and AO2 (application).		
Strengths of the answer			
Ways to improve the answer			
Level		**Mark**	

Examiner feedback

As with all 20-mark questions, there are four levels for this question. The style of the answer is a little muddled and rambling in places, but the candidate has created an answer in which there are some good, relevant points.

- The candidate has some fairly good knowledge and understanding (AO1) of glacial and periglacial conditions and landforms. Examples are given.
- The paragraphs are organised between glacial and periglacial conditions and landscapes, and then the difference between present and relict landscapes, all of which are located.
- To earn a higher mark, the candidate needs to develop more application to the question. There is a conclusion at the end but it needs to be built up more throughout the answer – it is rather bolted on to the last paragraph. Try to drop in small 'mini-conclusions' at the end of a paragraph before the main one at the end,

For AO1, the descriptor for Level 3 applies:

- *Knowledge and understanding is mostly clear, accurate and relevant to the argument, in explaining concepts and processes, and in explaining geographical scale and change over time.*

For AO2, Level 3 of the mark scheme applies to the first part:

- *Application is sound in different contexts (e.g. in different examples used), shows thorough, clear, and relevant analysis.*

However, the conclusion needs to be developed more throughout the answer, so Level 2 is more appropriate:

- *Leads to a generally sound, evaluative conclusion (though with little ongoing evaluation throughout the answer) which is sometimes based on evidence.*

The answer therefore fits the criteria for Level 3 overall, though not fully, because the conclusion is not well developed. Therefore an examiner would award this low-mid Level 3 for 13 marks.

4 Now mark this one!

Read through Sample Answer 2 on the next page.

a) Go through the answer using the three colours in section 3, and underline any links back to the question.
b) Look for evidence of both knowledge and understanding.
c) Use the mark scheme in section 1.7 to decide which level it is in and how many marks it is worth.

❓ Question recap

Question 4 Evaluate the statement that 'while glacial landscapes show ample evidence of global warming, it is actually periglacial landscapes that are most vulnerable to changes in climate'. (20 marks)

Glacial landscapes are those which are being actively modified, or have been glaciated in the past in upland and lowland regions. They are formed by ice masses, either in sheet form across wide expanses, or contained within upland regions as mountain and valley glaciers. Periglacial landscapes are also both active and relict, and consist of landforms formed around the margins of glaciers and ice sheets. Such regions are typified by tundra regions of, for example, Scandinavia. Both landscapes result from long-term changes to climate – both cooling during periods of glacial advance, and retreating as warming occurs. In the shorter term, climate change affects also both the size and movement of active glaciers and the occurrence of permafrost, both seasonally and over the medium term. Both sets of changes make glacial and periglacial regions vulnerable.

All glacial and periglacial landscapes change over time, simply through long-term erosional processes, and seasonal advance and retreat. As a result, there is a wide variety of both glacial and periglacial landscapes that are subject to change. Nepal is especially vulnerable to the impacts of climate change in mountain glaciers, with evidence that Himalayan glaciers are retreating rapidly. In 2013, researchers found that some glaciers around Mount Everest had shrunk by 13% since 1960 and that the snowline now is 180 metres higher than it was then. Satellite images suggest that glaciers are disappearing faster each year and that smaller glaciers are only half the size that they were in 1960. Retreat of this kind is leading to the growth of lakes dammed by glacial debris, which are then threatened by avalanches and earthquakes which breach them, causing catastrophic floods.

The degree of change is linked to the type of landscape and also to the scale and pace of climate change. There is widespread evidence now of global temperature increase, and that this is accelerating. The impacts of such change are hard to predict and both glacial and periglacial landscapes vary in the degree of change. However, it is clear that the highest and most sustained increases in temperature have occurred in polar and alpine environments. In polar regions, tundra margins are showing evidence of rapid thawing of permafrost in northern Sweden. This is making tundra ecology vulnerable, as short-rooted plants and low-lying shrubs are being out-shaded by tree growth which occurs when permafrost layers gradually thaw and permit deeper root growth.

Mountain landscapes are also just as vulnerable to change. Seasonal discharge causes retreat and loss from the glacial system and the result is a net loss of ice annually. This means that features such as outwash plains and moraines alter annually so, in many ways, these areas are more sensitive to changes in climate than lowland regions, especially plains, where relief is flatter, and surface run-off is reduced. However, seasonal thawing results in sudden increases in sediment yield which are associated with initial climate change. These are followed later by decay as margins progressively retreat from the periglacial landscape. Sediment change also occurs on slopes, where increased diurnal and seasonal freeze-thaw produces greater accumulation of scree, with associated mass movement such as rockslides and landslips on steeper slopes, and solifluction on lower and gentler slopes.

When glaciers melt, glacial mass balances are disrupted which, in turn, risks disrupting the hydrological cycle, and in the long term leads to reduced water supply for communities. The growth of glacial lakes is linked to glacial retreat. Moraine-dammed lakes are often located near glacier snouts. As glaciers melt, these lakes increase in size, and the moraine wall may collapse – causing a glacial lake outburst flood. In August 1985, one outburst flood caused a surge of water and debris down the Bhote Koshi river in Nepal, destroying a small HEP project.

Short-term climate changes affect relict glacial and periglacial landscapes much less. Relict features, from depositional features such as eskers, to erosional features such as striations, are modified as gradual warming occurs, rather than changing dramatically. In contrast, active periglacial and active glacial landscape regions are likely to evidence greater change. Evidence of glacier retreat is of interest but these changes show less vulnerability, whereas tundra ecosystems are lost as climates warm. Periglacial landscapes evidence far greater ecological change, whereas landform changes are less dramatic.

	Remember! Check for both AO1 (knowledge and understanding) and AO2 (application).		
Strengths of the answer			
Ways to improve the answer			
Level		**Mark**	

⑤ Marked sample answers

Sample Answer 2 is marked on the next page. The text has been highlighted to show how well each answer has structured points.

The following have been highlighted:

- points in red
- explanations in orange
- evidence in blue
- underlined points are where the candidate links to the question to identify the the extent to which periglacial landscapes are most vulnerable to change.

Look for the introduction and conclusion, which are essential for questions with the command word 'Evaluate'.

Marked sample answer 2

Link – the candidate introduces the answer well, defining both glacial and periglacial landscapes, and setting out the issue within the question

Glacial landscapes are those which are being actively modified, or have been glaciated in the past in upland and lowland regions. They are formed by ice masses, either in sheet form across wide expanses, or contained within upland regions as mountain and valley glaciers. Periglacial landscapes are also both active and relict, and consist of landforms formed around the margins of glaciers and ice sheets. Such regions are typified by tundra regions of, for example, Scandinavia. Both landscapes result from long-term changes to climate – both cooling during periods of glacial advance, and retreating as warming occurs. In the shorter term, climate change affects also both the size and movement of active glaciers and the occurrence of permafrost, both seasonally and over the medium term. Both sets of changes make glacial and periglacial regions vulnerable.

Point – refers first to changes in both types of landscape

All glacial and periglacial landscapes change over time, simply through long-term erosional processes, and seasonal advance and retreat. As a result, there is a wide variety of both glacial and periglacial landscapes that are subject to change. Nepal is especially vulnerable to the impacts of climate change in mountain glaciers, with evidence that Himalayan glaciers are retreating rapidly. In 2013, researchers found that some glaciers around Mount Everest had shrunk by 13% since 1960 and that the snowline now is 180 metres higher than it was then. Satellite images suggest that glaciers are disappearing faster each year and that smaller glaciers are only half the size that they were in 1960. Retreat of this kind is leading to the growth of lakes dammed by glacial debris, which are then threatened by avalanches and earthquakes which breach them, causing catastrophic floods.

Evidence – gives evidence of climate change and glacial retreat in Nepal

Explanation – extends the explanation and outlines impacts of changes

Point – the pace of change is discussed

The degree of change is linked to the type of landscape and also to the scale and pace of climate change. There is widespread evidence now of global temperature increase, and that this is accelerating. The impacts of such change are hard to predict and both glacial and periglacial landscapes vary in the degree of change. However, it is clear that the highest and most sustained increases in temperature have occurred in polar and alpine environments. In polar regions, tundra margins are showing evidence of rapid thawing of permafrost in northern Sweden. This is making tundra ecology vulnerable, as short-rooted plants and low-lying shrubs are being out-shaded by tree growth which occurs when permafrost layers gradually thaw and permit deeper root growth.

Evidence – gives evidence of the impacts of change on the tundra

Explanation – the candidate explains how change is affecting tundra landscapes

Point – the candidate switches from periglacial to mountain regions

Evidence – exemplars are given about specific changes to these regions; these are not named locations, but they don't need to be, it's the specific detail that's needed

Point – links are made to the hydrological cycle

Evidence – exemplars given from Nepal

Evidence – evidence given to illustrate the point

Link – provides a sound conclusion answering the wording of the question

Mountain landscapes are also just as vulnerable to change. Seasonal discharge causes retreat and loss from the glacial system and the result is a net loss of ice annually. This means that features such as outwash plains and moraines alter annually, so in many ways these areas are more sensitive to changes in climate than lowland regions, especially plains, where relief is flatter, and surface run-off is reduced. However, seasonal thawing results in sudden increases in sediment yield which are associated with initial climate change. These are followed later by decay as margins progressively retreat from the periglacial landscape. Sediment change also occurs on slopes, where increased diurnal and seasonal freeze-thaw produces greater accumulation of scree, with associated mass movement such as rockslides and landslips on steeper slopes, and solifluction on lower and gentler slopes.

When glaciers melt, glacial mass balances are disrupted which, in turn, risks disrupting the hydrological cycle, and in the long term leads to reduced water supply for communities. The growth of glacial lakes is linked to glacial retreat. Moraine-dammed lakes are often located near glacier snouts. As glaciers melt, these lakes increase in size, and the moraine wall may collapse – causing a glacial lake outburst flood. In August 1985, one outburst flood caused a surge of water and debris down the Bhote Koshi river in Nepal, destroying a small HEP project.

Short-term climate changes affect relict glacial and periglacial landscapes much less. Relict features, from depositional features such as eskers, to erosional features such as striations, are modified as gradual warming occurs, rather than changing dramatically. In contrast, active periglacial and active glacial landscape regions are likely to evidence greater change. Evidence of glacier retreat is of interest but these changes show less vulnerability, whereas tundra ecosystems are lost as climates warm. Periglacial landscapes evidence far greater ecological change, whereas landform changes are less dramatic.

Explanation – extends the impacts of changes in mountain regions

Explanation – these impacts are explained

Point – common features of glacial and periglacial landscapes established

Explanation – extended points about impacts of change are greatest in active periglacial and active glacial regions

---

✓ **Examiner feedback**

The descriptor for Level 4 applies to all parts:

- The candidate's knowledge of glacial landscapes and processes is wide ranging and accurate, and there is good use of examples. It meets all parts of the descriptor for Level 4 – i.e. knowledge and understanding which is detailed, thorough, relevant to the argument, accurate in explaining key concepts and processes, and has an awareness of geographical scale and change over time.

AO2 has all the qualities for Level 4, except that vulnerability is not discussed. Application is thorough in a range of examples given about glacial landscapes, and the analysis is detailed, coherent and relevant. There is a sound conclusion based on evidence.

By meeting the Level 4 descriptors, the answer earns 17 marks. To reach full marks, the candidate needs to have:

- discussed more about vulnerability – this is not defined but is an important part of the question. It ought to be mentioned fully in the conclusion.

On your marks
20-mark questions on Global systems and global governance

In this section you'll learn how to...

- maximise marks on 20-mark questions on Global systems and global governance, which use the command term 'To what extent?'.

Tackling 20-mark questions about Global systems and global governance

This topic is assessed on Paper 2, Section A, Question 1.

- Before beginning this section, you should study Section 1.7 in this book carefully. It will tell you about the skills you need to answer 20-mark questions, and the mark scheme.

Try this 20-mark question on Global systems and global governance

> This question assesses AO1 for 10 marks and AO2 for 10 marks.

Question

> To what extent can globalisation be said to have promoted economic growth and development, but also inequalities? (20 marks)

1 Plan your answer

Before attempting to answer the question, remember to **BUG** it. That means:

✓ **Box** the command word.
✓ **Underline** the following:
  - the theme
  - the focus
  - any evidence required
  - the number of points needed.
✓ **Glance** back over the question – to make sure you include everything in your answer.

Use the **BUG** on the next page to plan your own answer.

 Five steps to success!

Five steps to help you write top quality answers

The following five steps are used in this book to help you get the best marks.

1. **Plan your answer** – decide what to include and how to structure your answer.

2. **Write your answer** – use the answer spaces to write your answer

3. **Mark your answer** – use the mark scheme (section 1.7) to self- or peer-mark your answer. You can also use this to assess sample answers in step 4 below.

4. **Sample answers** – two sample answers are given to show you how to maximise marks for a question.

5. **Marked sample answer** – this is the same answer that you used for step 4 above, but is marked and annotated, so that you can compare with your own answers.

Command term: 'To what extent?' means use evidence to determine and weigh up the ways in which globalisation has created winners and/or losers within some countries, and reach a judgement.

Focus: The focus is 'globalisation' so you need to know something about the processes by which globalisation takes place, as well as some of its impacts.

To what extent can globalisation be said to have promoted economic growth and development, but also inequalities? (20 marks)

Evidence: 'economic growth and development' and 'inequalities' means that you must use evidence from specific examples (e.g. from developed or emerging economies) to develop and support an argument about ways in which globalisation has created benefits and problems.

Number of points: 20 marks. You need five or six reasoned points, preferably 2 or 3 on either side of the argument (AO2), supported by evidence (AO1).

PEEL your answer

Use PEEL notes to structure your answer. This will help to explain your ideas to the examiner clearly. PEEL has four stages:

- **P**oint – Make five or six developed points for 20 marks. These are the **AO2** parts of the answer where you argue about the impacts of globalisation in particular places.
- **E**vidence – Include details such as data or specific places to illustrate your argument. It's important to quote examples here, such as which countries have suffered greater inequalities because of globalisation. This is **AO1** – using knowledge to support your argument.
- **E**xplain – Give a reason for each point. Use starter phrases such as: '*Economic growth has taken place in the world's emerging countries because ...*'. This extends your **AO2** ability because you'll show how globalisation may be benefiting some countries more than others.
- **L**ink back to the question – Use the wording of the question in your answer to show you recognise what the question is asking, e.g. '*This example shows that the greatest inequalities from globalisation have occurred in ...*'.

 Tip

Remember, quality not quantity

You will not be marked on the **number** of points you make, but on the **quality** of your answer. That means the quality of the argument (AO2), the evidence you give (AO1), and how you link back to the question.

Remember the AOs!

It's important to know the AOs for this question.
- **AO1** (10 marks) is about demonstrating knowledge and understanding of globalisation, and of ways in which it has promoted economic growth, development, but also inequalities. Use brief examples – not long case studies.
- **AO2** (10 marks) is about using evidence to judge whether globalisation has promoted economic growth and development and also inequalities.

2 Write your answer

Using file paper, write your answer to the following question.

To what extent can globalisation be said to have promoted economic growth and development, but also inequalities? (20 marks)

3 Mark your answer

1. To help you to identify well-structured points, highlight or underline:

- points in red
- explanations in orange
- evidence in blue
- links back to the question by underlining.

Look for an introduction and conclusion, which are essential for questions with the command term 'To what extent?'

2. Use the mark scheme in section 1.7. Remember that:

- 20-mark questions are marked by choosing a level and a mark within it, based upon the answer as a whole
- high-scoring answers must include **both** AOs. One which includes only AO1 without discussing the winners and losers for AO2 can only gain a maximum of 10 marks.

Things to watch out for

Study Sample Answer 1 below.

- The candidate has done some things well, so identify which qualities are good.
- Identify what prevents the answer from getting a higher mark.

   Clues:
   Look at how the candidate demonstrates:
   a) **AO1** – what does the candidate know about globalisation? about economic growth and development? about inequalities?
   b) **AO2** – how well does the candidate develop arguments about globalisation and its impacts? Does the candidate use evidence to support an argument?

Remember the introduction and conclusion!

All 20-mark questions with the command term 'To what extent?' need to have:
- a brief (1–2 sentence) introduction
- a final short paragraph conclusion.

Globalisation means we are increasingly connected to other places with advances in transport and technology. This has led to advances in trade, tourism and migrations of people. It impacts upon different countries in different ways.

The heart of the global economy is where most wealth is produced in areas of North America, Europe, and Japan. These wealthy countries own and consume over 80% of global goods and services, and they earn the highest incomes. They also make most decisions about the global economy, e.g what goods are produced and where investment should be most directed in order to expand. Globalisation has been encouraged by TNCs who decide to manufacture overseas, where labour costs are less, so they can invest there. The TNCs are from the world's wealthy countries in North America, Japan, Europe, and wealthy investors from the Middle East.

Countries where they have invested have grown a lot – China, India and countries in south-east Asia have become manufacturing regions. India also provides a range of financial and IT support services for more developed countries. This is called outsourcing. The fashion industry makes its products in countries like Vietnam and Bangladesh which have grown fast. However, there are also the 'have-nots' in the world in sub-Saharan Africa and the world's poorest countries which have little economic influence in the world and are not sharing in growth. Countries in sub-Saharan Africa are less connected than countries in south-east Asia with fewer high-speed internet connections or infrastructure like airports of an international standard.

Some countries have low inequality due to globalisation. It means that incomes within a country do not vary a great deal and that people are more equal. The Netherlands has a Gini Index score of 27%, which is better than most other developed countries. The Netherlands also scores highly on the HDI index which shows that the population is doing well as a whole out of globalisation. But there are also some inequalities. Globalised countries have high internet usage, which is another good impact of globalisation. Internet usage in many developed countries is high, and is higher among younger people. Older generations are less likely to gain from this though. Advances in technology like this have benefited young people who can connect with anyone anywhere in the world.

Globalisation has also led to large increases in migrations of people. This is due to good transport all over the world. Part of this is due to countries which form trading blocs, within which people can move freely for jobs, e.g. in the EU. So countries like the UK and Netherlands have large ethnic populations and also Eastern European populations who have moved there for work. Most migrants end up living in large cities because that's where work is more plentiful. But they sometimes suffer racism and abuse.

	Remember! Check for both AO1 (knowledge and understanding) and AO2 (application).
Strengths of the answer	
Ways to improve the answer	

Level		**Mark**	

Examiner feedback

The candidate has put an argument together.

- Globalisation is defined – though the definition is not strong.
- There is some knowledge and understanding (AO1) of examples, e.g. about China's economic growth. The candidate also demonstrates knowledge about internet usage and migrations, and also about global economic growth.
- To earn a higher mark, the candidate needs to:
  a) define globalisation more clearly in a more detailed introduction
  b) develop more balance in the answer by explaining inequalities in more detail. For example, Bangladesh is shown as a country where economic growth has taken place, but there is no emphasis on low-paid work in sweatshops for many people. This candidate has painted economic growth as though it is all good.
  c) consider some of the downsides of migration, e.g. increased racism towards migrants, which is mentioned but not in any detail.

For AO1, the descriptor for Level 3 applies:

- *Knowledge and understanding* is mostly clear, accurate and relevant to the argument, in explaining concepts and processes, and in explaining geographical scale and change over time.

For AO2, Level 3 of the mark scheme also applies:

- *Application is sound in different contexts (e.g. in different examples used), shows thorough, clear, and relevant analysis*, and leads to a conclusion (with some ongoing evaluation throughout the answer) which is based on evidence.

The answer therefore fits the criteria for Level 3, though not fully, because the conclusion is not well developed. Therefore an examiner would award this low-mid Level 3 for 12 marks.

4 Now mark this one

Read through Sample Answer 2 opposite.

a) Go through the answer using the three colours in section 3, and underline any links back to the question to assess economic growth/development and inequalities.
b) Look for evidence of both knowledge and application.
c) Look for an introduction and conclusion, which are essential for questions with the command term 'To what extent?'.
d) Use the mark scheme in section 1.7 to decide which level it is in and how many marks it is worth.

? **Question recap**

To what extent can globalisation be said to have promoted economic growth and development, but also inequalities? (20 marks)

Globalisation means the ways in which people, culture, finance, goods and information move between countries with few barriers. Manufacturing industries have shifted to poorer countries where labour costs are low. This has led to a large increase in employment in shipping and trade as flows of raw materials and finance shift towards emerging countries. Globalisation also involves movements of people, capital, culture and technology. London, for example, attracts half of the world's capital investment through its financial industries, where people gain through highly paid work.

Growth and development have taken place in many Asian countries affected by globalisation. Since the 1970s, China's economy has doubled in size every eight years, while America's economy has doubled once. China is now an emerging economy and will soon be the world's largest economy. Its largest city, Shanghai, has benefited from major infrastructure projects and its people enjoyed waged work. Its office tower blocks give a sign of China's future, when manufacturing will have moved elsewhere and the 'new' knowledge and information sector take over. About a billion people in the world – half of them in China – have been lifted out of poverty since the 1990s, as a result of economic growth. So it seems inequality has been reduced while growth has taken place.

However, the impacts of globalisation have varied globally. One impact is on wealth distribution. Within a country, impacts depend on how evenly spread out the wealth of a country is, shown by the Gini coefficient. As an example, the Netherlands has a score of 27% which shows wealth is spread fairly equally. But at the same time 14% of the population of the Netherlands live below the poverty line, so globalisation has not benefited everyone. In some cases, the global shift to China and India has meant fewer jobs in countries like The Netherlands and more in China.

Globalisation involves companies trading globally, as well as governments of more developed countries becoming members of international organisations such as trade blocs (e.g. the EU) and IGOs (e.g. the UN). Some, like the EU, allow freedom of movement. Because it is easier to travel, people take more overseas holidays, travel for business or study abroad.

Improved communications mean that use of the internet increases to invest or purchase goods, or to communicate (e.g. Skype and social media). However, the impacts of globalisation can affect different generations in different ways, creating inequality. In countries in the EU, nearly 100% of those aged 16–24 have access to the internet, compared with 75% of those who are aged 55 or more. Internet usage is linked to connectivity. Those people who are more 'connected' and 'switched on' are more likely to have career advantages with jobs in the 'new economy' of quaternary and services, where salaries are often higher. But there are losers in successful countries. Those without internet or technology skills are more likely to end up in low-paid manual employment.

Other inequalities develop between cities and rural areas. Major differences exist in China between rural and urban dwellers in terms of employment type and incomes. People in urban areas in the coastal belt between Shanghai and Hong Kong earn much more than people left behind in rural areas. There are more people with migrant backgrounds in urban areas than rural, because of job availability. Therefore there are likely to be people with more global connections in urban areas who can send back remittance payments to those in rural areas, so that they win from globalisation as well. But the elderly and the poor are often left behind in rural areas, with a wide urban–rural gap.

Some countries welcome globalisation and play a large part in the global economy like Singapore. But sometimes the government stands in the way – for example, although China is a major global manufacturer, its population is less 'switched on' because the government uses firewalls to control internet activity as a way of preventing true freedom of speech. There have been protests in China against government control of freedoms but the government cracks down on these. So there are inequalities of freedoms even in countries like China which do very well economically.

	Remember! Check for both AO1 (knowledge and understanding) and AO2 (application).		
Strengths of the answer			
Ways to improve the answer			
Level		**Mark**	

⑤ Marked sample answers

Sample Answer 2 is marked opposite. The text has been highlighted to show how well each answer has structured points.

The following have been highlighted:

- points in red
- explanations in orange
- evidence in blue
- underlined points are where the candidate links to the question to identify the extent to which globalisation has created winners and losers.

Marked sample answer 2

Link – the candidate defines globalisation as a key term within the wording of the question

Globalisation means the ways in which people, culture, finance, goods and information move between countries with few barriers. Manufacturing industries have shifted to poorer countries where labour costs are low. This has led to a large increase in employment in shipping and trade as flows of raw materials and finance shift towards emerging countries. Globalisation also involves movements of people, capital, culture and technology. London, for example, attracts half of the world's capital investment through its financial industries, where people gain through highly paid work.

Evidence – a range of examples from China provide evidence of economic growth

Point – refers to growth and development as an impact of globalisation

Growth and development have taken place in many Asian countries affected by globalisation. Since the 1970s, China's economy has doubled in size every eight years, while America's economy has doubled once. China is now an emerging economy and will soon be the world's largest economy. Its largest city, Shanghai, has benefited from major infrastructure projects and its people enjoyed waged work. Its office tower blocks give a sign of China's future, when manufacturing will have moved elsewhere and the 'new' knowledge and information sector take over. About a billion people in the world – half of them in China – have been lifted out of poverty since the 1990s, as a result of economic growth. So it seems inequality has been reduced while growth has taken place.

Explanation – explains how inequality has been reduced.

Link – links inequality, growth and development back to the question

Point – makes a point about wealth distribution

However, the impacts of globalisation have varied globally. One impact is on wealth distribution. Within a country, impacts depend on how evenly spread out the wealth of a country is, shown by the Gini coefficient. As an example, the Netherlands has a score of 27% which shows wealth is spread fairly equally. But at the same time 14% of the population of the Netherlands live below the poverty line, so globalisation has not benefited everyone. In some cases, the global shift to China and India has meant fewer jobs in countries like The Netherlands and more in China.

Evidence – uses data about The Netherlands to support the point

Explanation – explains the point by referring to the shift in employment

Globalisation involves companies trading globally, as well as governments of more developed countries becoming members of international organisations such as trade blocs (e.g. the EU) and IGOs (e.g. the UN). Some, like the EU, allow freedom of movement. Because it is easier to travel, people take more overseas holidays, travel for business or study abroad.

Point – explains impact of trading blocs

Evidence – uses exemplar of EU freedom of movement

Explanation – extends the point using the example of greater travel freedom

Explanation – explains the connection between employment, connectedness, and the new economy

Improved communications mean that use of the internet increases to invest or purchase goods, or to communicate (e.g. Skype and social media). However, the impacts of globalisation can affect different generations in different ways, creating inequality. In countries in the EU, nearly 100% of those aged 16–24 have access to the internet, compared with 75% of those who are aged 55 or more. Internet usage is linked to connectivity. Those people who are more 'connected' and 'switched on' are more likely to have career advantages with jobs in the 'new economy' of quaternary and services, where salaries are often higher. But there are losers in successful countries. Those without internet or technology skills are more likely to end up in low-paid manual employment.

Point – changes direction to explain how globalisation affects generations differently

Evidence – uses data from the EU to illustrate the point

Point – explains urban–rural inequalities resulting from globalisation

Explanation – explains why urban–rural links are important

Other inequalities develop between cities and rural areas. Major differences exist in China between rural and urban dwellers in terms of employment type and incomes. People in urban areas in the coastal belt between Shanghai and Hong Kong earn much more than people left behind in rural areas. There are more people with migrant backgrounds in urban areas than rural, because of job availability. Therefore there are likely to be people with more global connections in urban areas who can send back remittance payments to those in rural areas, so that they win from globalisation as well. But the elderly and the poor are often left behind in rural areas, with a wide urban–rural gap.

Evidence – comparative data given for China

Link – links to a mini-conclusion about rural inequalities

Point – makes point about how governments may not welcome globalisation

Some countries welcome globalisation and play a large part in the global economy like Singapore. But sometimes the government stands in the way – for example, although China is a major global manufacturer, its population is less 'switched on' because the government uses firewalls to control internet activity as a way of preventing true freedom of speech. There have been protests in China against government control of freedoms but the government cracks down on these. So there are inequalities of freedoms even in countries like China which do very well economically.

Evidence – illustrates the point using China

Link – provides a mini-conclusion about inequalities of freedoms

Examiner feedback

This is a strong answer except for one factor!
The descriptor for Level 4 applies to all parts.

- The candidate's knowledge of globalisation and examples is wide ranging and accurate. It meets all parts of the descriptor for Level 4, i.e. knowledge and understanding is detailed, thorough, relevant to the argument, accurate in explaining key concepts and processes, and has an awareness of geographical scale and change over time.
- AO2 has all the qualities for Level 4 except one. Application is thorough in a range of examples given about globalisation, and the analysis is detailed, coherent and relevant.
- However, a final detailed evaluative conclusion is missing. It need only be a short 2–3 sentence paragraph, but it is needed. The links show evidence of ongoing evaluation throughout the answer, which is a good quality to have in an answer, but a final conclusion is needed to round it off well.

By meeting the Level 4 descriptors mostly, the answer earns 18 marks in the middle of Level 4. Had there been a conclusion, it might have earned the full 20 marks!

In this section you'll learn how to...

- maximise marks on 20-mark questions on Changing places, which use the command term 'How far do you agree?'.

Tackling 20-mark questions about Changing places

This topic is assessed on Paper 2, Section B, Question 2.

- Before beginning this section, you should study section 1.7 carefully which will tell you about the skills you need to answer 20-mark questions, and the mark scheme.

Try this 20-mark question on Changing places

> This question assesses AO1 for 10 marks and AO2 for 10 marks.

Question

How far do you agree that the policies of external agencies such as governments impact upon people and places? (20 marks)

① Plan your answer

Before attempting to answer the question, remember to **BUG** it. That means:

✓ **Box** the command word.
✓ **Underline** the following:
  - the theme
  - the focus
  - any evidence required
  - the number of points needed.
✓ **Glance** back over the question – to make sure you include everything in your answer.

Use the **BUG** on the next page to plan your own answer.

Five steps to success!

Five steps to help you write top quality answers

The following five steps are used in this book to help you get the best marks.

1. **Plan your answer** – decide what to include and how to structure your answer.

2. **Write your answer** – use the answer spaces to write your answer

3. **Mark your answer** – use the mark scheme (section 1.7) to self- or peer-mark your answer. You can also use this to assess sample answers in step 4 below.

4. **Sample answers** – two sample answers are given to show you how to maximise marks for a question.

5. **Marked sample answer** – this is the same answer that you used for step 4 above, but is marked and annotated, so that you can compare with your own answers.

Command term: How far do you agree? means recognising that policies of external agencies can impact upon people and places on a spectrum from large to small, and deciding where on that spectrum you'd place them.

Evidence: How far you agree that 'policies of external agencies ... impact upon people and places' means that you should use evidence of policies to decide your judgement based on examples you have studied.

How far do you agree that the policies of external agencies such as governments impact upon people and places? (20 marks)

Focus: The focus is about the characteristics and impacts of external forces operating upon places, specifically external forces such as government policies, TNCs or international and global institutions.

Number of points: 20 marks. You need five or six reasoned points, preferably balanced across both sides of the argument (AO2), supported by evidence (AO1).

PEEL your answer

Use PEEL notes to structure your answer. This will help to explain your ideas to the examiner clearly. PEEL has four stages:

- **P**oint – Make five or six developed points for 20 marks. These are the **AO2** parts of the answer where you discuss the impacts of attempts by local and national government decision-makers to change places, and show how far you agree with the statement.
- **E**vidence – Include details about specific ways in which attempts by local and national government decision-makers to change places illustrate your argument. It's important to use examples, such as whether they are short- or long-term, or economic or social changes. This is **AO1** – using knowledge to support your argument.
- **E**xplain – Give a reason for each point. Use starter phrases such as: '*This particular role of local government is important because ...*'. This extends your **AO2** ability because you'll show that you can identify the significance of particular changes.
- **L**ink back to the question – Use the wording of the question in your answer to show you recognise what it is asking, e.g. '*National government has played a large role in this particular change because ...*'.

 Tip

Remember, quality not quantity

You will not be marked on the **number** of points you make, but on the **quality** of your answer. That means the quality of the argument (AO2), the evidence you give (AO1), and how you link back to the question.

Remember the AOs!

It's important to know the AOs for this question.
- **AO1** (10 marks) is about demonstrating knowledge and understanding of policies of external agencies such as governments to bring changes to places. Use brief examples – not long case studies.
- **AO2** (10 marks) is about using evidence to decide how far you agree that policies of external agencies such as governments have brought change, and how.

② Write your answer

Using file paper, write your answer to the following question.

Question

How far do you agree that the policies of external agencies such as governments impact upon people and places? (20 marks)

③ Mark your answer

1. To help you to identify well-structured points, highlight or underline:

- points in red
- explanations in orange
- evidence in blue
- links back to the question by underlining.

Look for an introduction and conclusion, which are essential for questions with the command term 'How far do you agree?'.

2. Use the mark scheme in section 1.7. Remember that:

- 20-mark questions are marked by choosing a level and a mark within it, based upon the answer as a whole
- a high-scoring answer must include **both** AOs. One which includes only AO1 without applying the question for AO2 would gain a maximum of 10 marks.

Things to watch out for

Study Sample Answer 1 opposite.

- The candidate has done some things well, so identify which qualities are good.
- Identify what prevents the answer from getting a higher mark.

    Clues:
    Look at how the candidate demonstrates:
    a) **AO1** – what does the candidate know about changes to places and the part played by external agencies such as governments in bringing about these changes?
    b) **AO2** – how well does the candidate develop arguments about the part(s) played by external agencies such as governments in bringing about change? Are particular changes assessed for the part played by governments?

Remember the introduction and conclusion!

All 20-mark questions with the command term 'How far do you agree? need to have:
- a brief (1–2 sentence) introduction
- a final short paragraph conclusion.

Both local and national government decision-makers are important in bringing about changes to places. Other players also play key roles, e.g. housing associations and local community groups, and architects.

Government involvement in changing places is often driven by a policy of bringing economic growth and improving quality of life. Glasgow is a good example of a city which required change. In Glasgow the national government used data about levels of multiple deprivation to identify which parts of the city needed changing most. Local councils were responsible for implementing changes.

The Scottish government provided some funding and also decided on the marketing and rebranding of parts of the city to attract new investors. The biggest project to change the city was the 2014 Commonwealth Games. The Games had a global TV audience of over 1 billion. The national government spent £750m on the Games, which created many jobs in the building sector and gave Glasgow some much improved sports facilities. So the national government was important in funding this.

Other changes were brought by the city council in Glasgow. One project was the Glasgow Arc, costing £20m. The Arc is a bridge which has been designed to improve the look of the city. It looks good lit up at night and so makes a safe environment around the area. Another large project was the development of a financial quarter by the city council. It has attracted big companies such as JP Morgan, as well as a £188m BBC Scotland headquarters. Much of the funding came from national government, but the decision-making and planning was done by the city council. Since it opened, the financial district has attracted £1bn investment which has created more jobs.

The city council also decided how to reduce council housing stock in Glasgow by transferring 80 000 council houses to the Glasgow Housing Association (GHA) – this was one of the biggest housing transfers in UK history. It reduced the total council housing stock in Glasgow from 90% to 50%. Since the transfer, housing associations have spent over £1.5bn in improving tenants' houses. Other organisations have been responsible for bottom-up projects such as the Castlemilk Housing Cooperative responsible for improving buildings in Glasgow (led by residents).

Architects have also been important decision-makers in changing Glasgow – there are good and bad examples of this. During Glasgow's first regeneration projects, Sir Basil Spence's poor design of council tower blocks showed how poor architecture can create poor standards of living – the flats were damp, badly looked after, and hotspots of drug crime. But more recent architectural innovations, such as the Armadillo and the Riverside Museum, have helped Glasgow establish itself a national reputation as a rebranded city.

Government decision makers are vital to city regeneration projects and change. Local government has set up projects, and national government helps in funding and rebranding the city for things like the Commonwealth Games and City of Culture.

	Remember! Check for both AO1 (knowledge and understanding) and AO2 (application).		
Strengths of the answer			
Ways to improve the answer			
Level		**Mark**	

✓ **Examiner feedback**

As with all 20-mark questions, there are four levels for this question. The candidate has created some relevant points.

- The candidate has very good knowledge and understanding (AO1) of changes to Glasgow, and of who funds and initiates projects.
- There is some application (AO2) – the candidate refers to national and local government in different projects, though goes beyond this by referring to architects and others. It would help the candidate to clarify which people were funded by national or local government, as appropriate.
- To earn a higher mark, the candidate needs to apply more thought to the question. The roles of different governments are identified, but are not compared with other organisations. Therefore, the conclusion is weak.

For AO1, the descriptor for Level 4 applies:

- The answer shows detailed *knowledge and understanding* which is thorough, and relevant, accurate and shows awareness of geographical scale and change over time.

For AO2, Level 3 mostly applies:

- *Application is sound in different contexts*
- *Analysis is thorough, clear, and relevant*
- *The conclusion is clear and evaluative* which is based on evidence.

With Level 4 for AO1, and fitting most of the criteria at Level 3 for AO2 (except for the last criterion), the answer needs a 'best fit'. In this case, markers would award the middle part of Level 3 overall, i.e. 13 marks.

4 Now mark this one!

Read through Sample Answer 2 below.

a) Go through the answer using the three colours in section 3, and underline any links back to the question.
b) Look for evidence of both knowledge and application.
c) Look for an introduction and conclusion, which are essential for questions with the command term 'How far do you agree?'.
d) Use the mark scheme in section 1.7 to decide which level it is in and how many marks it is worth.

 Question recap

How far do you agree that the policies of external agencies such as governments impact upon people and places? (20 marks)

Sample Answer 2

Changes to places include modifications which arise from a wish to encourage economic growth, or to improve quality of life. Both local and national government decision-makers are important in the regeneration of urban areas. The private sector often plays a role, e.g. banks or property developers, and architects often breathe new thinking and reuses of land. But their work is difficult without either national or local government approval. This essay will assess the role of local and national government in order to consider their impact.

National governments bring significant change to places by investing in infrastructure. At the moment, this includes HS2 and Heathrow Airport's third runway. Both are expensive and a profitable return is unlikely in the short term, so attracting private sector investment is difficult. National government argues that improvements such as these are essential to economic growth by improving accessibility, and that these in turn encourage further investment and spur regeneration through the economic multiplier effect. Once built, local government may benefit and drive further change. But national government is the initial driver.

National governments also see a broader picture from a national perspective. Sometimes, greater funding is needed for events which are considered to be in the national interest, and where local interests have failed to bring about change. For example, London's 2012 Olympic Games were supported by the London Mayor, and born out of wishes to reduce the wealth gap between London's wealthiest and poorest parts (e.g. east London). The barrier was cost – £9 billion – and it is here that national government supplied greater funding with the will to implement it. The Games were considered to be of national benefit, whilst locally east London gained from the 'spirit' of the Games, the subsequent investment in Stratford and environmental improvement in the Olympic Park.

Nonetheless, local government plays its part, which may be increasing because of national decisions to hand greater power to local assemblies and councils. Some major UK cities, such as London and Manchester, operate almost as a mini states (e.g. the Mayor of London's office) and can change processes affecting places at an appropriate level. London's transport network has been revolutionised by the congestion charge, which has generated income for public transport, resulting in a doubling in bus passenger numbers since 2003, extensions to the tube network and DLR, and the introduction of night trains at weekends. Change may have been driven by national government, but it is local government which creates these changes.

Other areas in which national policies affect changes to places include migration. National government makes decisions over international migration and attracting foreign direct investment to the UK. But it is at local government level that these policies focus on local need, and local councils decide on whether to give planning permission for development, for example a factory. This can have a backlash – in the 1980s, local borough councils in east London were against Docklands regeneration by the Thatcher government, so that local planning powers were removed and handed to governmental Urban Development Corporations. But in most cases, local councils are the decision-makers about changes.

In conclusion, governments make laws and set policies affecting places (e.g. house building, fracking, buy-to-let rules). They may also make grants that affect economic and environmental regeneration in urban areas. But the implementation of policies, and the reasons why London and Manchester often adopt different solutions to similar problems (e.g. transport), is down to local councils, often influenced by interest groups such as the Chamber of Commerce or even pressure groups, as in the case of fracking. Local councils are frequently swayed by local opinion which may influence decisions. Cuts to national government spending (e.g. the abolition of regional development agencies in 2010) have often reduced local options, but other funding helps (e.g. National Lottery funding for a range of local projects). So it is external agencies that often change places in a big way.

	Remember! Check for both AO1 (knowledge and understanding) and AO2 (application).
Strengths of the answer	
Ways to improve the answer	

Level		**Mark**	

5 Marked sample answers

Sample Answer 2 is marked below. The text has been highlighted to show how well each answer has structured points.

The following have been highlighted:

- points in red
- explanations in orange
- evidence in blue
- underlined points are where the candidate links to the question to identify the impacts of the policies of external agencies.

Marked sample answer 2

Link – the candidate introduces the answer well, using the wording of the question, and defines both key terms as well as setting out the context

Changes to places include modifications which arise from a wish to encourage economic growth, or to improve quality of life. Both local and national government decision-makers are important in the regeneration of urban areas. The private sector often plays a role, e.g. banks or property developers, and architects often breathe new thinking and re-uses of land. But their work is difficult without either national or local government approval. This essay will assess the role of local and national government in order to consider their impact.

Point – the candidate refers to the role of national government in infrastructure

National governments bring significant economic change to places by investing in infrastructure. At the moment, this includes HS2 and Heathrow Airport's third runway. Both are expensive and a profitable return is unlikely in the short term, so attracting private sector investment is difficult. National government argues that improvements such as these are essential to economic growth by improving accessibility, and that these in turn encourage further investment and spur regeneration through the economic multiplier effect. Once built, local government may benefit and drive further change. But national government is the initial driver.

Evidence – gives examples of infrastructure changes

Explanation – the candidate explains the impact of investment and the multiplier effect

Link – the candidate links the point to the question – an example of ongoing evaluation

National governments also see a broader picture from a national perspective. Sometimes, greater funding is needed for events which are considered to be in the national interest, and where local interests have failed to bring about change. For example, London's 2012 Games were supported by the London Mayor, and born out of wishes to reduce the wealth gap between London's wealthiest and poorest parts (e.g. east London). The barrier was cost – £9 billion – and it is here that national government supplied greater funding with the will to implement it. The Games were considered to be of national benefit, whilst locally east London gained from the 'spirit' of the Games, the subsequent investment in Stratford and environmental improvement in the Olympic Park.

Point – the candidate refers to the situation when national government takes control

Evidence – uses the example of the 2012 Olympic Games

Explanation – the candidate extends the impacts of the 2012 Games investment

Point – the candidate switches focus from national to local government

Explanation – compares the role of national and local government

Link – the comparison links directly back to the question

Evidence – gives an exemplar of Docklands in the 1980s

Point – begins the conclusion by setting out the key role of national government

Links – a good distinction is made between local and national government roles

Evidence – gives examples of how national government spending cuts have reduced role of local government

Nonetheless, local government plays its part, which may be increasing because of national decisions to hand greater power to local assemblies and councils. Some major UK cities, such as London and Manchester, operate almost as a mini states (e.g. the Mayor of London's office) and can change processes affecting places at an appropriate level. London's transport network has been revolutionised by the congestion charge, which has generated income for public transport, resulting in a doubling in bus passenger numbers since 2003, extensions to the tube network and DLR, and the introduction of night trains at weekends. Change may have been driven by national government, but it is local government which creates these changes.

Other areas in which national policies affect changes to places include migration. National government makes decisions over international migration and attracting foreign direct investment to the UK. But it is at local government level that these policies focus on local need, and local councils decide on whether to give planning permission for development, for example a factory. This can have a backlash – in the 1980s, local borough councils in east London were against Docklands regeneration by the Thatcher government, so that local planning powers were removed and handed to governmental Urban Development Corporations. But in most cases, local councils are the decision-makers about changes.

In conclusion, governments make laws and set policies affecting places (e.g. house building, fracking, buy-to-let rules). They may also make grants that affect economic and environmental regeneration in urban areas. But the implementation of policies, and the reasons why London and Manchester often adopt different solutions to similar problems (e.g. transport), is down to local councils. often influenced by interest groups such as Chambers of Commerce or even pressure groups, as in the case of fracking. Local councils are frequently swayed by local opinion which may influence decisions. Cuts to national government spending (e.g. the abolition of regional development agencies in 2010) have often reduced local options, but other funding helps (e.g. National Lottery funding for a range of local projects.) So it is external agencies that often change places in a big way.

Evidence – exemplars are given about the role of local urban government and transport

Point – migration referred to as a national policy

Explanation – distinguishes between national and local government

Link – refers the point back to the question

Explanation – extends the point further

Evidence – uses examples of London and Manchester

Point – a further distinction made about the role of local government

Link – the candidate concludes with an answer to the question

Examiner feedback

This is a strong answer. The descriptor for Level 4 applies to all parts:

- The candidate's knowledge is wide ranging and accurate about economic changes. The answer shows *knowledge and understanding* which is detailed, thorough, relevant, accurate in explaining key concepts and processes and shows careful awareness of geographical scale and change over time. (AO1)
- The answer is thorough in a range of *contexts and scales* (local, regional and national issues are discussed), while *analysis* is detailed, coherent and relevant, leading to an evaluative conclusion, as well as ongoing *evaluation* throughout the answer, which is rational and firmly based on evidence. (AO2)

By meeting the Level 4 descriptors fully, the answer earns 18 marks. To reach full marks, the candidate needs to have strengthened the final conclusion.

Geographical skills

Introduction

AS and A Level Geography requires students to demonstrate a variety of geographical skills, showing a critical awareness of the appropriateness and limitations of different methods, skills and techniques.

The aim of this chapter is to explain, illustrate and allow you to practise the key skills central to both AS and A Level specifications.

Many will have been developed throughout GCSE studies and even before. Furthermore, new skills will have been integrated into the course content rather than taught separately. Consequently, don't be alarmed if you cannot recall specific lessons on the seemingly wide range of skills covered in the exam specification. Note also that many geographical skills (in the broadest sense of the phrase) are those invaluable 'working' skills essential to the well-informed, enquiring and competent adult. The geography 'tag' merely reinforces why this discipline above all others is central to any good, 'rounded' education.

Look at Figure **2**. Think about these key verbs (found in the Geographical skills checklist in section 3.4 of the specification):

* understanding
* collecting
* identifying
* questioning
* interpreting
* analysing
* evaluating
* applying
* communicating
* arguing.

You will have been developing competence and confidence in these elements throughout the course – especially during fieldwork investigation. In short, you'll have been continually reinforcing and honing these working skills – not just in geography, but selectively in other subjects too.

"Who needs to learn Geography anymore when we have Google Earth?!"

▲ *Figure 1*

Key definitions

Qualitative data is descriptive. It is exploratory in nature, involving in-depth research and analysis, such as data collected in one-to-one interviews as part of a fieldwork investigation.

Quantitative data is measurable. It is numerical and so can be verified and transformed into usable statistics.

Primary data is collected first-hand, such as during fieldwork. Consequently, it is real-time data specific to the needs of the investigation.

Secondary data is collected by others, and often used to support the primary data or to allow studies of changes over time. Census data and other information collected by government departments are good examples.

'Big data' describes extremely large datasets often requiring huge amounts of computational power. It is often used to reveal patterns, trends, and associations, and so very much associated with 'geolocation' and the geospatial data stored, displayed and analysed using Geographic Information Systems (GIS).

Understand the nature and use of different type of geographical information, including:
* qualitative and quantitative data
* primary and secondary data
* images, factual text and discursive/creative material
* digital data, numerical and spatial data
* innovative forms of data, including crowd-sourced and 'big data'.

Collect, analyse and **interpret** such information, and demonstrate the ability to understand and **apply** suitable analytical approaches for the different information types.

Geographical skills

Undertake informed and critical **questioning** of data sources, analytical methodologies, data reporting and presentation, including the ability to **identify** sources of error in data and to identify the misuse of data.

Communicate and **evaluate** findings, draw well-evidenced conclusions informed by wider theory, and construct extended written **argument** about geographical matters.

▲ *Figure 2* AS and A Level Geographical skills

The specification (section 3.4.2) then goes on to identify the specific qualitative and quantitative skills you need to cover (including ICT skills familiar to you, but not covered in this book). Page numbers in this book are given in red:

AS and A Level Geographical skills (Specification reference in brackets)				
Core (3.4.2.1)	**Cartographic (3.4.2.2)**	**Graphical (3.4.2.3)**	**Statistical (3.4.2.4)**	**ICT (3.4.2.5)**
Literacy – use of text *Throughout book*	Atlas maps *pp120–123*	Line graphs – simple, comparative, compound and divergent *pp140–143*	Measures of central tendency – mean, mode, median *pp160–163*	Use of remotely sensed data (as described in Core skills)
Numeracy – use of number *Throughout book*	Maps with located proportional symbols *pp120–123*	Pie charts and proportional divided circles *pp121, 123, 140–143*		
Questionnaire and interview techniques	Weather maps *pp124–127*	Bar graphs – simple, comparative, compound and divergent *pp140–143*	Measures of dispersion – range, interquartile range and standard deviation *pp 160–163, 164–167*	Use of electronic databases
OS maps (at a variety of scales) *pp132–135*	Maps showing movement – flow lines, desire lines and trip lines *pp128–131*	Scattergraphs, and the use of best fit line *pp144–147*	Spearman rank correlation coefficient test *pp168–171*	Use of innovative sources of data such as crowd sourcing and 'big data'
Use and annotation (inc. overlays) of illustrative and visual material – maps, diagrams, graphs, field sketches and photographs *pp136–139*	Maps showing spatial patterns – choropleth and dot maps *pp120–123*	Triangular graphs *pp148–151*	Chi-squared test *pp172–175*	Use of ICT to generate evidence of skills such as producing maps, graphs and statistical calculations
		Graphs with logarithmic scales *pp156–159*	Application of significance tests *pp169–175*	
Use of geospatial, geolocated and digital imagery (see ITC skills)	Maps showing spatial patterns – isoline maps *pp124-127*	Dispersion diagrams *pp160–163*		

🔺 **Figure 3** *Skills identified in the specification 3.4.2*

Finally, don't forget that all skills are primarily a useful, supportive tool. Revisit and practise them as you would geographical theory, and maintain a 'can do' attitude. Whether you are adopting a cartographic, graphical or statistical skill, keep your objective (purpose) clearly in mind and:

- remember that informing the reader (examiner) is your focus
- be measured, but not pedantic
- check any calculations as you progress
- review again following other tasks – especially at the end of the exam.

3.1 Choropleth maps, dot maps and proportional circles

You need to know:

- the comparative strengths and weaknesses of choropleth and dot maps
- how to use proportional circles effectively.

Used effectively, **choropleth** and **dot maps** and also **proportional circles** are excellent ways of showing spatial distributions and variations across areas.

But an understanding of their strengths and weaknesses are required in order to avoid adopting the wrong format, or misrepresenting data patterns through poor technique and misinterpreting the map information in subsequent analysis.

Choropleth maps

Choropleth maps are atlas and textbook favourites for showing global patterns, such as population trends (e.g. density, life expectancy), variations in wealth (e.g. GDP and GNP per capita) and comparative standards of living (e.g. HDI, health care). They show these patterns in a gradation of colour and/or density shadings (Figure 1). Maps showing regional variations may also use choropleth shading.

Ideally, there should be between four and six categories, with eight as an absolute maximum. Category values should not overlap, preferably be equal or, if not, be in a logical geometric sequence.

A drawback in their use is that they imply uniformity, by averaging values across a wide area (e.g. county or country), and therefore hide local variations. Furthermore, the sudden changes at boundaries are highly unlikely in reality.

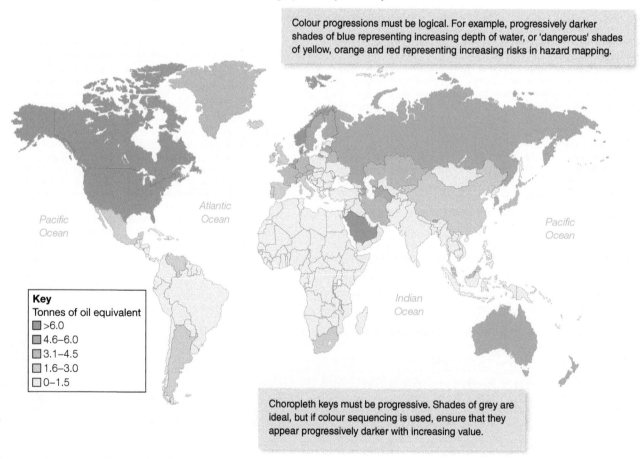

Colour progressions must be logical. For example, progressively darker shades of blue representing increasing depth of water, or 'dangerous' shades of yellow, orange and red representing increasing risks in hazard mapping.

Key
Tonnes of oil equivalent
- >6.0
- 4.6–6.0
- 3.1–4.5
- 1.6–3.0
- 0–1.5

Choropleth keys must be progressive. Shades of grey are ideal, but if colour sequencing is used, ensure that they appear progressively darker with increasing value.

⬥ **Figure 1** *Choropleth map showing energy consumption per capita (2014)*

Dot maps

When dot maps are executed well, they can leave powerful impressions. Located dots are used to represent a particular value or number, with their density (spread) creating the visual impact (Figure **2**).

But both careful choice of dot value and their size on the map are critical in order to avoid creating misleading impressions (e.g. areas devoid of people in a population distribution map). Their greatest strength is immediacy of impact, particularly should dots merge. However, they are difficult to extract accurate information from.

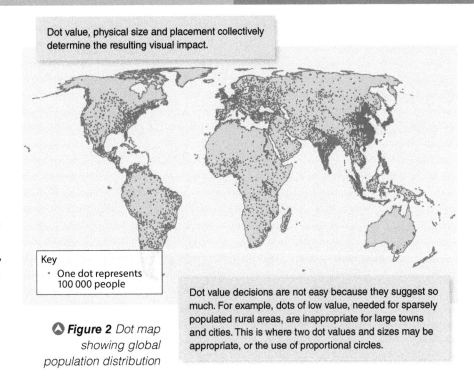

Dot value, physical size and placement collectively determine the resulting visual impact.

Key
· One dot represents 100 000 people

Dot value decisions are not easy because they suggest so much. For example, dots of low value, needed for sparsely populated rural areas, are inappropriate for large towns and cities. This is where two dot values and sizes may be appropriate, or the use of proportional circles.

🔺 **Figure 2** *Dot map showing global population distribution*

Proportional circles

Proportional circles (and to a lesser degree proportional squares) are very useful for comparing located data (e.g. city populations or factory production values). The size of each circle is proportional to the value it represents. Given thoughtful consideration of scale, and careful adherence to good practice, proportional circle maps can be very informative (Figure **3**).

Constructing proportional circles

* The area of the circle should be proportional – so the square roots of the values to be plotted could simply be each circle's radius.
* Check this for your biggest value and go ahead if the circle looks sensible on your map.
* If not, decide the size you want your largest circle to be.
* Calculate the diameter of the remaining circles using the following formula:

$$\text{circle size} = \text{maximum circle size} \times \left(\frac{\text{the sq. root of the value}}{\text{the maximum sq. root value}} \right)$$

* Plot your circles and add a key (ranging from smallest to largest).

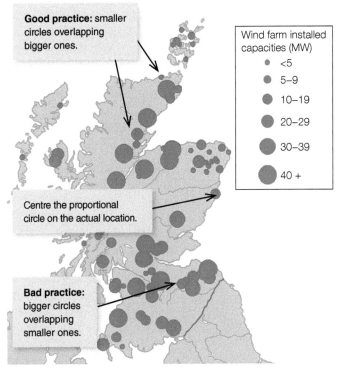

Good practice: smaller circles overlapping bigger ones.

Wind farm installed capacities (MW)
· <5
· 5–9
· 10–19
· 20–29
· 30–39
· 40 +

Centre the proportional circle on the actual location.

Bad practice: bigger circles overlapping smaller ones.

🔺 **Figure 3** *Proportional circles map showing Scottish onshore wind farm capacities*

Sixty second summary

* Choropleth and dot maps can be excellent ways of showing spatial distributions and variations across areas.
* Choropleth maps use grades of colour and/or density shadings to show grouped values.

* Dots representing particular values, located on a map, have great impact in showing density (spread).
* Proportional circles (with the size of each circle in proportion to the value it represents) are very useful for comparing located data.

Now practise...

Question 1

Study Figure **4**.

Critically evaluate both the mapping technique adopted and the effectiveness of this map. **(9 marks)**

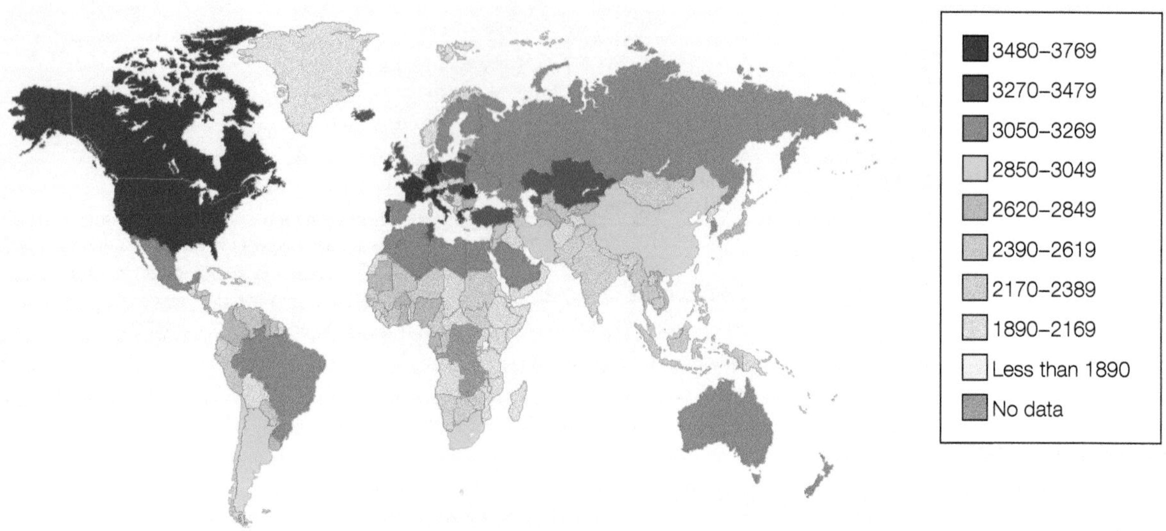

▲ *Figure 4* Global average daily calorie intake per capita (2011)

Legend:
- 3480–3769
- 3270–3479
- 3050–3269
- 2850–3049
- 2620–2849
- 2390–2619
- 2170–2389
- 1890–2169
- Less than 1890
- No data

_____

		Mark/9
Strengths of the answer		
Ways to improve the answer		

Question 2

Study Figure **5**.

Ski resort	Ten-year average visitor numbers	Square root of ten-year average	Diameter of circle (cm)
Cairngorm	67 391	259.60	1.5
Nevis Range	20 889	144.53	
Glencoe	17 061	130.62	
Glenshee	52 177		
Lecht	38 582		

⬥ **Figure 5** *Ten-year average visitor numbers for Scottish ski resorts*

a) Complete the table. (2 marks)

b) Plot the data as proportional circles on the map outline below. (6 marks)

c) Add a title and key for the proportional circles. (1 mark)

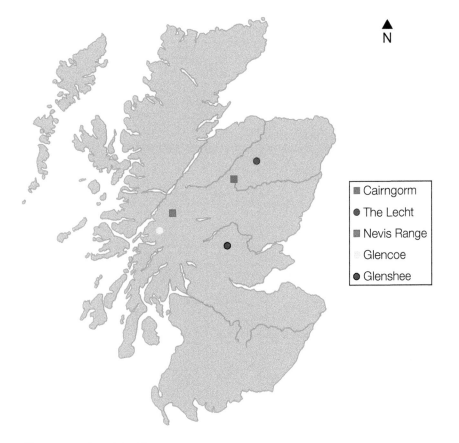

⬥ **Figure 6** *Scottish ski resorts*

		Mark/9
Strengths of the answer		
Ways to improve the answer		

You need to know:

- the variety, use and value of isolines on maps and diagrams.

What are isolines?

'Iso' means 'equal'. So isolines are 'lines of equal' variables on maps and diagrams used in a variety of situations. You are already likely to be familiar with their use. For example:

- **Isobars** (lines of equal barometric/air pressure), isotherms (lines of equal temperature) and isohyets (lines of equal precipitation) on weather and climate maps.
- **Contours** (lines of equal elevation or height) on relief maps.
- **Isovels** (lines of equal velocity) on river cross-sections (Figure **1**).

Why use isolines?

The value of isolines lies in their capacity to show located trends. They can be enhanced further by adopting choropleth shading (see 3.1), such as layer shading on relief maps. Plotting isolines requires a skill involving *interpolation* – a 'trial and error' process which can be mastered surprisingly quickly. Just as cartographers originally plotted relief on maps by interpolating contours on base maps showing surveyed dot heights, so you may find great benefit in the technique for plotting located trends in fieldwork data. For example:

- pedestrian and traffic densities in a CBD
- environmental quality variations across a wider urban area
- percentage vegetation cover on a psammosere (inland from beach sand dunes to climax vegetation).

 Tip

Plotting isolines
Follow some simple rules when plotting isolines:
- Never go 'straight to best' – it is crucial to prepare a rough draft to decide and check value intervals, and to judge if the resulting isolines 'look plausible'.
- Plot in sequence from biggest to smallest, or vice versa.
- Isolines should always flow in natural curves.
- Isolines should never cross.
- Remember that interpolation means 'estimating between known values' but is, in effect, interpretation. Consequently, your plotting will be unique.

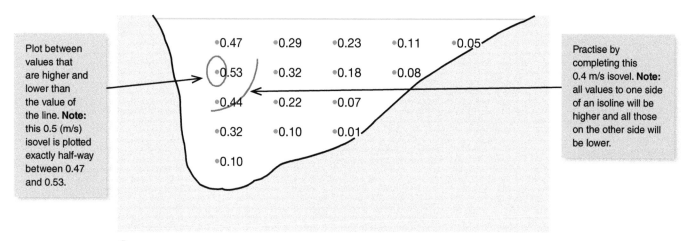

Plot between values that are higher and lower than the value of the line. **Note:** this 0.5 (m/s) isovel is plotted exactly half-way between 0.47 and 0.53.

•0.47   •0.29   •0.23   •0.11   •0.05

•0.53   •0.32   •0.18   •0.08

•0.44   •0.22   •0.07

•0.32   •0.10   •0.01

•0.10

Practise by completing this 0.4 m/s isovel. **Note:** all values to one side of an isoline will be higher and all those on the other side will be lower.

▲ **Figure 1** *Velocities (in m/s) at various depths across a river cross-section (profile)*

 Sixty second summary

- 'Iso' means 'equal'; so isolines are 'lines of equal' variables on maps and diagrams.
- Isolines join points of equal values (e.g. isotherms showings temperatures on a weather map).
- Isolines show located trends and can be enhanced further by adding choropleth shading.
- Isolines should always flow in natural curves and never cross.

Now practise …

Question 1

Study Figure **2**.

a) Plot isovels at intervals of 0.10 m/s. **(10 marks)**

b) Annotate the profile using the information on the plan view and your own knowledge. **(6 marks)**

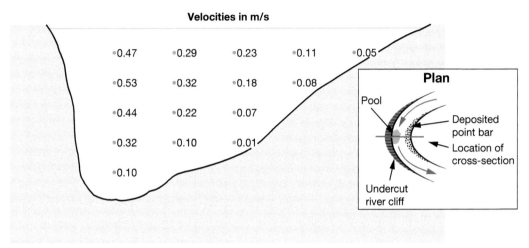

Velocities in m/s

•0.47   •0.29   •0.23   •0.11   •0.05

•0.53   •0.32   •0.18   •0.08

•0.44   •0.22   •0.07

•0.32   •0.10   •0.01

•0.10

Plan

Pool

Deposited point bar

Location of cross-section

Undercut river cliff

⬥ **Figure 2** *Velocities (in m/s) recorded using a flow vane at various depths across a river cross-section (profile)*

c) Explain the distribution of velocity shown in the channel cross-section. **(4 marks)**

_____

_____

_____

_____

_____

_____

_____

_____

_____

		Mark/20
Strengths of the answer		
Ways to improve the answer		

Question 2

Study maps **A** and **B** in Figure **3**.

a) On Map **B** plot isolines for 600 mm moisture loss due to evapotranspiration. **(6 marks)**

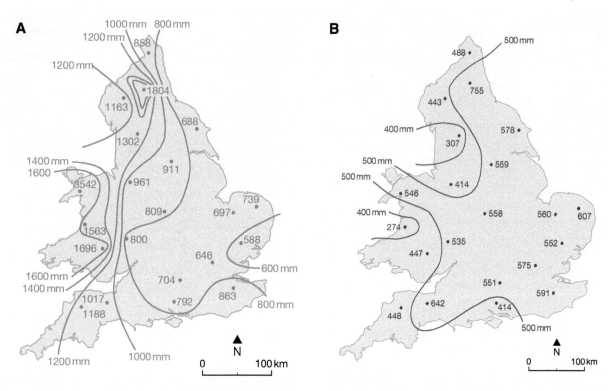

⬢ **Figure 3** *Maps showing (**A**) precipitation (mm) and (**B**) moisture loss due to evapotranspiration (mm) for England and Wales*

b) (i) Describe the broad relationships between precipitation (Map **A**) and moisture loss (Map **B**). **(6 marks)**

_____

_____

_____

_____

_____

_____

_____

_____

_____

_____

_____

_____

_____

_____

_____

(ii) State which parts of England and Wales are most likely to be affected by moisture deficiency?　　**(2 marks)**

_____

_____

_____

_____

c) Assess to what extent agricultural land-use trends are likely to be influenced by this relationship between precipitation and moisture loss due to evapotranspiration.　　**(6 marks)**

_____

_____

_____

_____

_____

_____

_____

_____

_____

_____

_____

_____

_____

		Mark/20
Strengths of the answer		
Ways to improve the answer		

You need to know:

- how to show the relative strength of interactions between variables on diagrams
- how to map the direction and volume of people or goods from one place to another.

Flow lines

Flow lines are a useful means of showing the relative strength of interactions between variables (Figure **1**) and especially the direction and volume of people or goods from one place to another along a specific route.

Proportional arrows (with the width drawn in proportion to the value being shown) are especially accessible and have an immediate impact. This is why they are widely adopted in graphics to show, for example:

- migration flows, such as so-called illegal immigrant crossings into Europe (Figure **2**)
- traffic movements at a local scale plotted on maps showing the actual routes
- trade in petroleum, such as crude oil exports from the Middle East (Figure **3**).

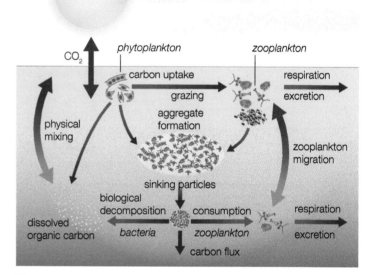

▲ *Figure 1 The biological carbon pump*

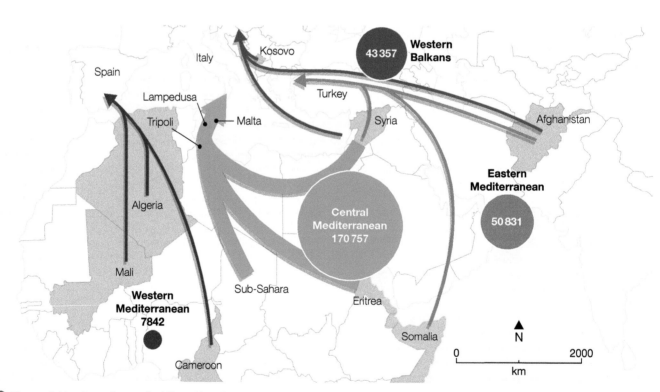

▲ *Figure 2 Number of so-called illegal immigrant crossings into Europe (2014). Note the use of proportional circles (see 3.1) to clearly show additional regional information in this complex, but informative map.*

Desire lines and trip lines

Desire lines are similar to flow lines but show only *direct* movement from A to B (and not the actual route). Consequently, in effect, they symbolise the movement (or flow if drawn proportionally). But you will find inconsistencies in different definitions. For example:

- some insist that desire lines should radiate from a single origin
- some state that they must be straight lines
- others allow sweeping curves (Figure **4**).

Similar inconsistencies are found when making the distinction between these and **trip lines**. These distinctions range from subtle to contradictory, so can very easily confuse – rather as with the variations in terminology for radial charts (see 4.5).

The most logical distinction is that a *desire line* shows average routes or paths used over time (Figure **4**) whereas a *trip line* marks a specific journey (and so could well be exactly the same). But don't be surprised when you find alternatives to this so, in practice, adopt whatever term the question demands. Whether 'desire line' or 'trip line', the key point to ensure is that you are accurate in your plotting of origin and destination.

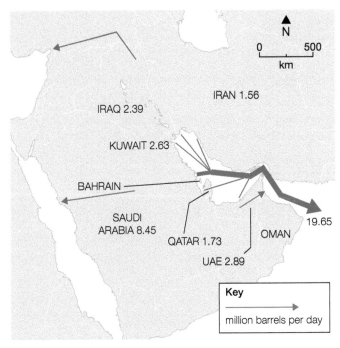

▲ **Figure 3** *Middle East oil exports (2013)*

▲ **Figure 4** *Desire line map of international flights from Heathrow, London*

> **Tip**
>
> **Plotting flow lines**
> Judgement is required during rough drafting of flow lines in order to ensure that the scale you choose is appropriate to clearly show both the smallest values (the thinnest arrow) and the largest (the thickest).
>
> Plotting from origin (source) to destination is simplest using straight lines which follow the exact path of movement.
>
> Try to avoid flow lines crossing over each other and don't forget to write the scale on your map.

Sixty second summary

- Flow lines on diagrams are a useful means of showing the relative strength of interactions between variables.
- Flow lines on maps are most commonly used to show the direction and volume of people or goods moving from one place to another along a specific route.
- Flow lines use arrows to show direction of movement and different widths (drawn in proportion to the value being shown) to show volumes.
- Desire lines and trip lines are similar to flow lines but show only direct movement from the place of origin to the destination (and not the actual route).

Now practise...

'*There are three kinds of lies: lies, damned lies, and statistics.*'

This infamous statement is as widely quoted as it is misquoted – and misattributed! (It was popularised in the USA by Mark Twain among others, who attributed it to the British Prime Minister Benjamin Disraeli.) Its relevance to this exercise will become clear.

Study the data in Figure **5**, which shows the number of National Insurance registrations in the UK by EU and non-EU nationals.

EU	NI Registrations (000s)	% change from 2016
Romania	154	−19%
Poland	62	−34%
Italy	51	−19%
Bulgaria	39	−8%
Spain	36	−25%

Non EU	NI Registrations (000s)	% change from 2016
India	32	−10%
Pakistan	12	0%
Australia	11	−12%
China	11	−15%
United States	10	−2%

Figure 5 *Top 5 individual nationalities (EU and non EU) for National Insurance number registrations (2017)*

a) Plot the National Insurance number registrations data (2017) from Figure **5** as flow lines on the map below. Choose a suitable scale (e.g. 1 mm = 20 000 registrations). **(10 marks)**

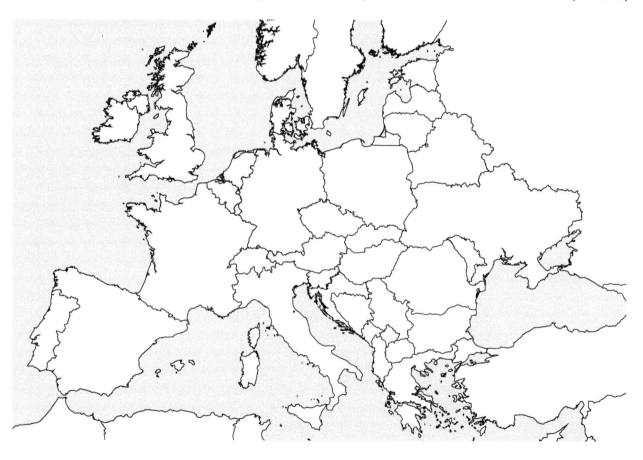

b) Explain why presenting the number of registrations as a flow map could be misleading.   **(4 marks)**

_____

_____

_____

_____

_____

_____

_____

_____

c) Explain **two** contrasting ways that the data in Figure **5** might be used to deliberately misinform.   **(6 marks)**

_____

_____

_____

_____

_____

_____

_____

_____

_____

_____

_____

_____

_____

		Mark/20
Strengths of the answer		
Ways to improve the answer		

You need to know:
- key OS map-reading skills, including grid references, scales and distance measurement
- how to identify and describe both physical and human features.

Key OS map reading skills

We use Ordnance Survey maps to locate places, determine area, distance and direction, and both visualise and understand land use, communications, relief and drainage (Figure **1**). Each map extract is a section of a national grid covering the whole of Great Britain.

Scales of 1:10 000, 1:25 000 and 1:50 000 are most commonly used. For example, a scale of 1:10 000 means that 1 unit on the map (e.g. 1 cm) represents 10 000 units on the ground (e.g. 10 000 cm).

Four- and six-figure grid references

All OS maps are covered by a numbered grid made up of sequential numbers – **eastings** (identifying longitude) and **northings** (identifying latitude). They form the basis of the **four-** and **six-figure references** used to locate features on the map. Four-figure references identify the whole square. Six-figure references locate specific points (by estimating tenths of the whole square – Figure **2**).

The number combinations work from the bottom left-hand corner, reading eastings first, then northings. (You 'crawl before you climb', so 'along the corridor and up the stairs'!)

Calculating distance and areas

Straight-line distances can be marked on the edge of a piece of paper and read directly from the map's 'linear' scale line. Curving distances, such as along a road, river or coastline are best measured by dividing the route into straight sections (Figure **3**).

Calculating areas requires judgement of the proportion of a grid square (or squares) an area feature occupies. On 1:25 000 and 1:50 000 maps, one square on the map represents 1 km² on the ground.

The 16-point compass

Always check the north point on a map; do not assume it is straight up (as on OS maps). Note that the top of OS maps point to *grid north* which is nearly, but not exactly the same as *true north*. (Your compass points to *magnetic north*, which is not true north either!)

In an 8-point compass the four main cardinal points (north, east, south and west) are divided by intercardinal points (NE, SE, SW and NW). A 16-point compass divides the cardinal and intercardinal points (e.g. ENE between E and NE).

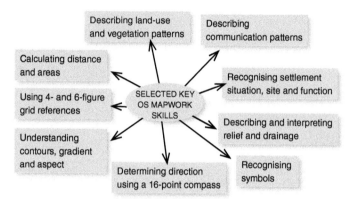

▲ **Figure 1** *Selected key OS mapwork skills*

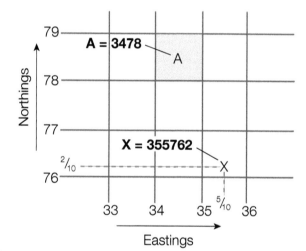

▲ **Figure 2** *Using 4- and 6-figure grid references*

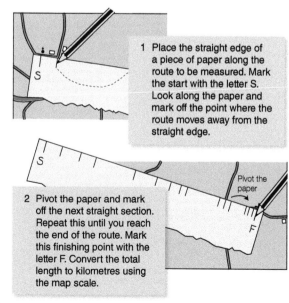

1 Place the straight edge of a piece of paper along the route to be measured. Mark the start with the letter S. Look along the paper and mark off the point where the route moves away from the straight edge.

2 Pivot the paper and mark off the next straight section. Repeat this until you reach the end of the route. Mark this finishing point with the letter F. Convert the total length to kilometres using the map scale.

▲ **Figure 3** *Measuring distance along a curved line*

Describing relief and drainage

Relief refers to the height and shape of the land. It can be interpreted on the map from contour patterns and supporting spot heights and triangulation ('trig') points. Contours are (brown/orange) lines of equal height (see 3.2) plotted at regular intervals – they are therefore widely spaced on gentle **gradients**, and closer together on steeper slopes.

Drainage refers to how water is drained from the land – including its flow and storage. For example, a dense pattern of river channels might indicate impermeable geology and/or high precipitation. Unnaturally straight channels and reservoirs indicate human intervention.

The study of contour and river patterns allows landscape visualisation that becomes second nature with practice (Figure **4**). Providing your description and explanation are supported by hard evidence from the map (e.g. specific heights, names, distances, directions faced (**aspect**), grid references) then your map interpretation will have authority.

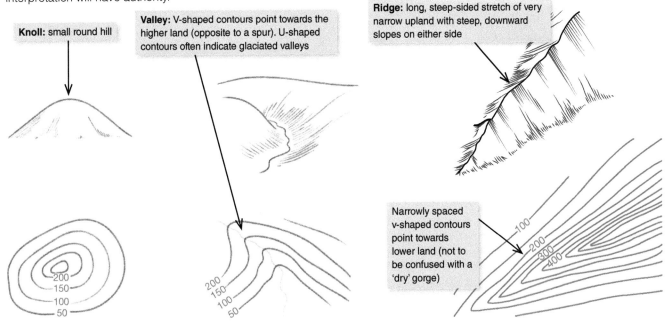

Knoll: small round hill

Valley: V-shaped contours point towards the higher land (opposite to a spur). U-shaped contours often indicate glaciated valleys

Ridge: long, steep-sided stretch of very narrow upland with steep, downward slopes on either side

Narrowly spaced v-shaped contours point towards lower land (not to be confused with a 'dry' gorge)

▲ **Figure 4** *Knoll, valley and ridge and their contour patterns*

Describing settlements, communications and land use

Describing settlement patterns involves terms such as **linear** (the built-up area extending along a road or river), **nucleated** (dense and focused on a central point) or **dispersed** (spread out at low density). But beware the common misuse of the terms **situation** and **site**. The former refers to the settlement's location relative to other places, whereas the latter describes the actual land it occupies (think of a building site and you won't go wrong). **Function** refers to principal characteristics such as 'market town', or 'heavy industrial centre'.

Communications primarily refer to transport networks, such as roads and railways, and frequently reflect the relief of an area. Again, description and explanation must be supported by specific, located evidence from the map.

Finally **land use** is self-evident, but may require reference to human modification or management. As always, reference to location is crucial, including, ideally, the size and shape of the area.

Sixty second summary

- Ordnance Survey maps are respected for accuracy, detail and intuitive symbols.
- The four-figure grid reference identifies the bottom left-hand corner of the square.
- Eastings are read before northings.
- On 1:25 000 and 1:50 000 OS maps, one square on the map represents 1 km² in real life.
- Always check the direction of the north point on a map.
- Directions may be given using the 8-point compass, but a 16-point compass allows for greater precision.
- Descriptions of relief and drainage patterns should always be supported by located references to specific heights, names, distances, aspect, density and so on.
- Descriptions and explanations of settlements, communications and land use must always be supported by specific, located evidence from the map.

Now practise...

Question 1

Study the OS map in Figure **5**. You will find the key to map symbols on page 176.

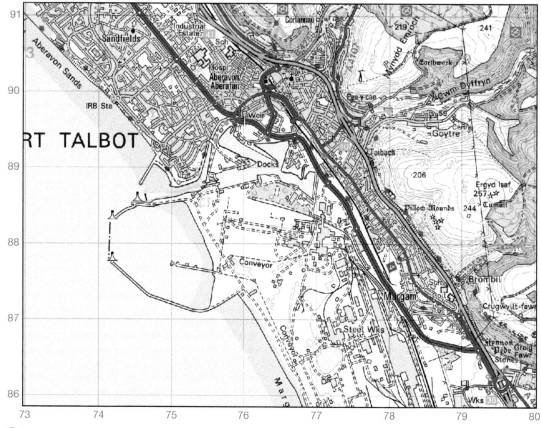

🔺 **Figure 5** *Port Talbot, OS Landranger 1:50000 series (extract). Note that one grid square is 1 km².*

    a) State the 6-figure grid reference for Sandfields church. _____ **(1 mark)**

    b) Estimate the area of sea at low water visible on this extract. _____ **(1 mark)**

    _____

    c) State the distance to, and direction of, the nearest beacon from the one at 742 883. **(2 marks)**

    _____

    _____

    d) Describe the gradient and aspect of the slope from 778 889 to 790 889. **(2 marks)**

    _____

    _____

    e) Describe the situation, site and probable function of Margam (78 87). **(4 marks)**

    _____

    _____

f) Describe the principal communications across this extract. **(4 marks)**

g) Contrast the relief, drainage and land use south of northing 89. **(6 marks)**

		Mark/20
Strengths of the answer		
Ways to improve the answer		

You need to know:

- how to sketch, label and annotate effectively.

Why sketch when we have photographs?

Photographs are enormously valuable to geographers. Whether used on their own, or in association with maps, they provide an instant visual reference, aiding investigation and understanding of both physical and human landscapes. But they must have a purpose beyond gratuitous decoration.

You should already be familiar with describing photographs and the use of directional language (e.g. 'in the background') and juxtaposition (e.g. 'just to the left of the statue') in order to make sense of what is being shown.

So why sketch when labelling and/or annotating photographs can have such an impact and value (Figure **1**)?

Sketching – whether field sketching on site or subsequently from a photograph – enables you to:

- take stock and think about what you are aiming to achieve
- focus on the subject – forms and processes
- identify the main geographical characteristics
- omit irrelevant detail.

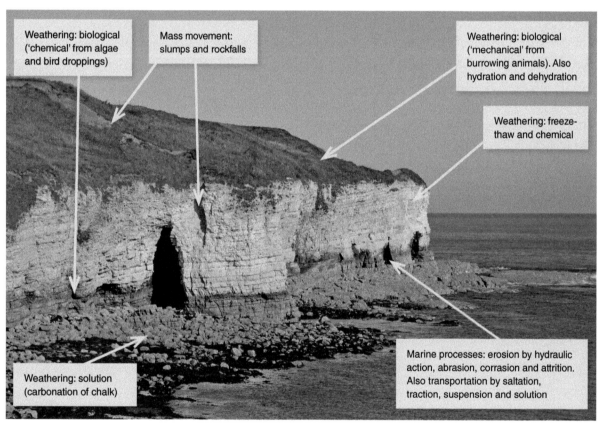

Weathering: biological ('chemical' from algae and bird droppings)

Mass movement: slumps and rockfalls

Weathering: biological ('mechanical' from burrowing animals). Also hydration and dehydration

Weathering: freeze-thaw and chemical

Weathering: solution (carbonation of chalk)

Marine processes: erosion by hydraulic action, abrasion, corrasion and attrition. Also transportation by saltation, traction, suspension and solution

Figure 1 *Marine and sub-aerial processes, Flamborough Head, East Yorkshire*

Drawing and labelling sketches

Whether sketching in the field or from a photograph, you don't have to be artistic. But you do need to be patient, clear and accurate so that your labelling and annotation makes sense – think about what you are trying to achieve. Also, if field sketching on site, don't forget to record the location (e.g. place name or grid reference) and from which direction the subject has been viewed.

Build your sketch in sequence (Figure **2**). Think about the all-important labels (identifying features) and annotation (detailed description and explanation) while you are sketching, but they can be added as the final stage.

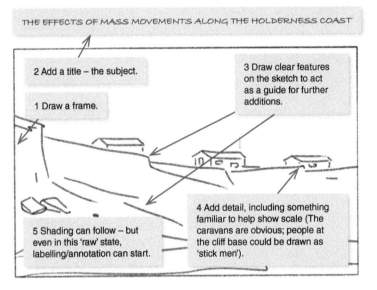

THE EFFECTS OF MASS MOVEMENTS ALONG THE HOLDERNESS COAST

2 Add a title – the subject.

1 Draw a frame.

3 Draw clear features on the sketch to act as a guide for further additions.

4 Add detail, including something familiar to help show scale (The caravans are obvious; people at the cliff base could be drawn as 'stick men').

5 Shading can follow – but even in this 'raw' state, labelling/annotation can start.

THE EFFECTS OF MASS MOVEMENTS ALONG THE HOLDERNESS COAST

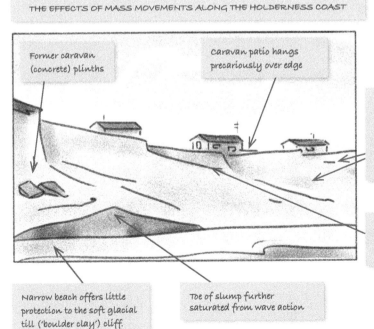

Former caravan (concrete) plinths

Caravan patio hangs precariously over edge

Slump debris – thriving grass suggests recent mass movements

Slip plane of rotational slip (slump)

Narrow beach offers little protection to the soft glacial till ('boulder clay') cliff. Longshore drift removes material down the coast

Toe of slump further saturated from wave action

Figure 2 *Sketching in the field or from a photo*

Tip

Drawing effective field sketches
- Secure your paper on a flat surface (e.g. clipboard)
- Decide whether your sketch will be landscape or portrait, depending on the view.
- Use plain paper (not lined) or faint graph paper if less confident.
- Use a pencil – you can correct as you progress.
- Hold your pencil vertically or horizontally at arm's length to work out the proportions. Sliding your thumb up and down allows key heights and widths to be reproduced accurately on your sketch. (Practise this by sketching an object on your desk.)

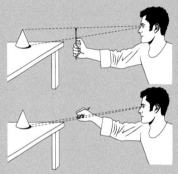

- Sketch horizons first, then outlines of the main features.
- Keep everything simple and neat.
- Once happy with the overall proportions start adding the important details.
- Only add shading if you are confident – don't overdo it.
- Make sure that you include a familiar object, animal or person to show scale.
- Label your sketch noting down carefully any measurements you have taken.

Sixty second summary

- Photographs provide an instant visual reference aiding investigation and understanding of both physical and human landscapes.
- Sketching enables you to be more discriminating – to take stock and to think about what you are aiming to achieve.
- Effective sketching in the field, or from photographs, allows you to identify key geographical characteristics, omit irrelevant detail, and focus on forms and processes.

Now practise...

In April and May 1991, around 30 million cubic metres of rock fell from the mountainsides near the tourist resort of Zermatt, levelling the railway and road, and blocking the Matter Vispa river (Figure **3**).

Although there were no fatalities, communications to the head of the valley and Zermatt were cut off, and the river soon began to flood behind the debris. Swiss army conscripts were mobilised and within four days, the road to Zermatt had been diverted around the debris, and within six weeks the rerouted railway line reopened.

A specific cause for these rockfalls remains the subject of debate. However, there are a number of possible explanations:

- The Swiss Alps remain tectonically active and a small (albeit unrecorded) earthquake could have been a contributory factor.
- Weathered soft clay or parallel planes of weakness (known as cleavage) may have finally been overcome by the weight of the rocks above.
- Isolated permafrost that cemented the rocks together may have melted – possibly as a result of global warming.

Study Figure **3**.

⊘ *Figure 3* *A 2011 photo of the 1991 Randa rockfalls, Valais, Switzerland*

Draw an annotated sketch to illustrate key geographical characteristics of the rockfalls, two decades after the events.

(9 marks)

		Mark/9
Strengths of the answer		
Ways to improve the answer		

- how to construct and interpret bar graphs, histograms, divided bar graphs, line graphs, compound line graphs and pie charts.

Why graph?

As a geographer, you need to be able to assess raw numerical data and then select an appropriate way to *graph* it so that the information is presented visually in a clear, informative and logical manner.

When constructing a graph, use a sharp pencil and double-check that you have plotted the data correctly. Keep things clear and simple – avoid overloading the graph with additional information (such as raw statistics) and fussy shading (such as hatching and cross-hatching).

Bar graphs, histograms and divided bar graphs

Bar graphs are commonly used to show data that is unconnected, such as quantities or frequencies in different categories (Figure **1**). They are conventionally drawn with different colours and a small gap between each bar.

Histograms are a type of bar graph showing data derived from a single population, sample or variable (Figure **2**). They are drawn using the same colour and the bars touching (both of which illustrate the connection). **Age-sex (population) pyramids** are a common form of 'mirrored' histogram showing the proportions of a population in different age and gender categories.

Divided bar graphs show multiple data by subdividing the individual bars. They are particularly useful when comparisons are required across place and/or time (Figure **3**).

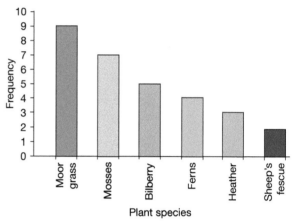

▲ **Figure 1** A bar graph showing different types of vegetation at a fieldwork location

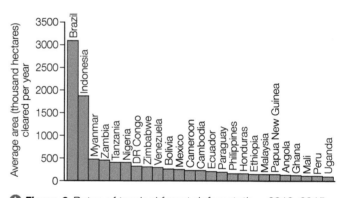

▲ **Figure 2** Rates of tropical forest deforestation, 2010–2015

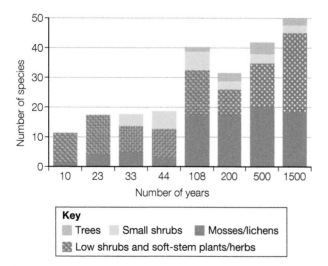

Key
- Trees
- Small shrubs
- Mosses/lichens
- Low shrubs and soft-stem plants/herbs

▲ **Figure 3** A divided bar graph showing the number of species exposed by a retreating glacier

 Tip

Describing and explaining graphs
When describing graphs, **trends**, **examples** and **anomalies** (**TEA**) should be identified:
- **T**rends cover the general direction – use adjectives such as rising, falling, steady, accelerating, flared.
- **E**xamples illustrate the general trend and should include the highest and lowest values.
- **A**nomalies should cover any exceptions to the general trend – such as examples that stand out as different, outliers/residuals in scattergraphs, radical changes of gradient in line or bar graphs.

When explaining graphs, give reasons for the *trends*, *patterns* and *anomalies* observed. Also consider the links between different variables that might help in the explanation.

Line graphs

Line graphs are used to show continuous data, usually showing changes taking place over time. It is possible to sub-divide the area below a line graph to show different proportions of the total. This creates a **compound line graph** (Figure **4**).

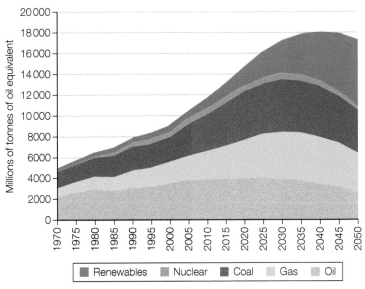

△ **Figure 4** Long-term global energy demand predictions

Pie charts

Pie charts show proportions of a total as segments of a circle. They work best when kept simple – between four and six segments, and using solid, contrasting colours (Figure **5**). However, annotation using raw percentage figures can help interpretation.

Percentages are converted into degrees for the pie chart by multiplying the value by 3.6.

Proportional pie charts are used when comparing two or more sets of data when the categories are similar but there is a change in another variable, such as time or absolute total (see 3.1).

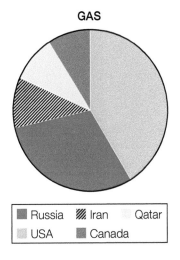

△ **Figure 5** The five largest natural gas producers (2015)

⏱ **Sixty second summary**

- Graphs show numerical information visually – in a clear, informative and logical manner.
- Bar graphs commonly show data that is unconnected, such as quantities or frequencies in different categories.
- Histograms are a type of bar graph showing data derived from a single population, sample or variable.
- An age–sex (population) pyramid is a histogram showing the proportions of a population in different age and gender categories.
- Divided bar graphs show multiple data by subdividing the individual bars.
- Line graphs show continuous data. They usually show changes taking place over time. Compound line graphs sub-divide the area below to show different proportions of the total.
- Pie charts show proportions of a total as segments of a circle.

Now practise...

Tabulation of raw numerical data has great value for identifying specific values. But geographers need to be able to visualise and understand trends, patterns and links in data, interpret mapping (see 3.1) and be able to graph data. These are invaluable skills.

Software tools have value in this respect. But their 'instant fix' lack the finesse and flexibility required for making sure that they are always fit for purpose. Hence the importance of learning how to construct and interpret appropriate maps and graphs.

'Choice means control' and this may well involve an element of creativity. For example, *composite graphs* have enormous potential, but must be constructed with care, so as not to overcomplicate.

 Tip

Ensuring accurate data representation
- Use a ruler to guide your eye.
- When using multiple scales, cover opposite scales (not used) with a strip of paper to stop your eye drifting to the nearest (incorrect) scale when plotting in the latter stages.
- If *x*-axis divisions are extended (e.g. months), plot line graph points centrally.
- Think carefully about what colours provoke what assumptions (e.g. red suggesting heat).

'Climographs' in atlases demonstrate this principle well. They superimpose line graphs (temperature) and histograms (precipitation) with matched scales to allow easy comparison.

How far you can develop composite graphs becomes a matter of judgement, but more than four scales are likely to clutter, especially if annotation is added to help you memorise detail.

Study Figure **6**.

Month	Maximum temp. (°C)	Minimum temp. (°C)	Rainfall (mm)	Relative humidity (%)
January	31	24	249	89
February	31	24	231	89
March	31	24	262	89
April	31	24	221	90
May	31	24	170	89
June	31	24	84	87
July	32	24	58	87
August	33	24	38	85
September	33	24	46	84
October	33	24	107	85
November	33	24	142	86
December	32	24	203	88

🔺 **Figure 6** *Climate data for Manaus, Brazil*

a) Using the frame and scales in Figure **7** as a guide, use the graph paper opposite to plot:

- rainfall as a histogram
- maximum, minimum and average temperatures as line graphs
- relative humidity as a line graph. **(10 marks)**

b) Annotate your composite graph with key identifying statistics and characteristics. **(8 marks)**

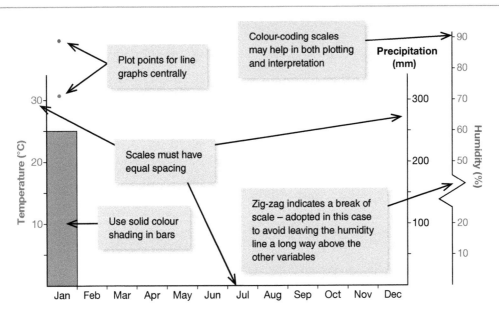

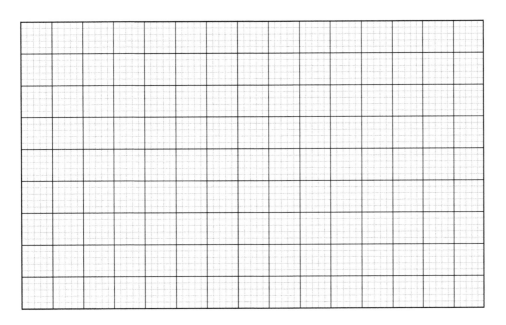

▼ **Figure 7** *Layout of climate graph for Manaus, Brazil*

c) Suggest whether or not there is scope to add further climate data to your graph. **(2 marks)**

_____

_____

_____

_____

		Mark/20
Strengths of the answer		
Ways to improve the answer		

Scattergraphs and trend lines

If two sets of data (variables) are thought to be related they can be plotted on a **scattergraph**.

Convention is to plot the 'independent variable' on the *x*-axis and the 'dependent variable' (the one you are seeking to understand) on the *y*-axis. Once plotted, the 'scattered' points may allow you to visualise a trend (correlation) or pattern.

Look at Figure **1**. Whether smooth curves or straight lines, trend lines should be drawn approximately through the middle of the points, with roughly the same number of points on either side. They do not need to pass through the origin.

A **trend line** of 'best-fit', however, is straight and drawn after calculating the mean of each data set, and plotting this as 'the mean point' (see 5.1). Rotating a ruler's edge around this point, and aligning with the trend, allows drawing of the best-fit line to show the correlation.

Whether this correlation is positive or negative, strong or weak will be estimated by the closeness of the plotted points to the best-fit line. But remember, outliers or 'residuals' lying well beyond the best-fit line should be ignored when plotting, but they should be examined in subsequent analysis (Figure **1**).

In some instances, a scattergraph pattern makes it impossible to draw a trend line. This would infer that there is no correlation. But statistical tests, such as the *product moment correlation*, and especially the *Spearman rank correlation coefficient* test (see 5.3), allow you to test whether or not a relationship does in fact exist, including the degree of confidence that it has not occurred by chance.

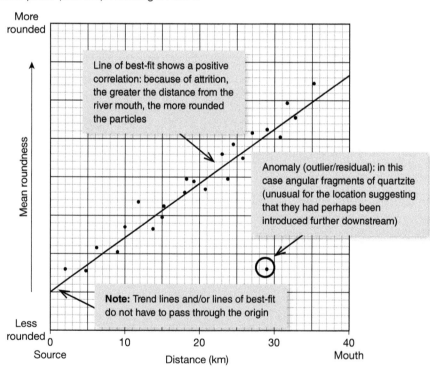

⬆ **Figure 1** *Changes in river particle roundness with distance from the source*

Now practise...

Question 1

Study Figure **2** and use the graph paper on the following page.

Country	GDP per capita (PPP) ($)	Infant mortality (rate per thousand)
Australia	49 900	4.3
Bangladesh	4200	31.7
Brazil	15 500	17.5
Chile	24 600	6.6
Cuba	11 900	4.4
Germany	50 200	3.4
India	7200	39.1
Japan	42 700	2.0
Kenya	3500	37.1
Mexico	19 500	11.6
Portugal	30 300	4.3
Switzerland	61 400	3.6

▲ **Figure 2** *The relationship between GDP per capita (PPP*) and infant mortality rates per thousand for selected countries in 2017*

*GDP per capita (PPP) figures are based upon purchasing power parity. They accurately reflect the economic condition of countries as a whole, but hide inequalities of wealth within countries.

 Tip

Describing graphs
When describing graphs (including scattergraphs), **trends**, **examples** and **anomalies** (**TEA**) should be identified:
- **T**rends cover the general direction – use adjectives such as rising, falling, steady, accelerating, flared.
- **E**xamples illustrate the general trend and should include the highest and lowest values.
- **A**nomalies should cover any exceptions to the general trend – such as examples that stand out as different, outliers/residuals in scattergraphs, radical changes of gradient in line or bar graphs.

a) (i) Draw a frame and suitable scales before plotting a scattergraph of the data. **(5 marks)**

   (ii) Draw a trend line. **(2 marks)**

   (iii) Identify any residual(s). **(2 marks)**

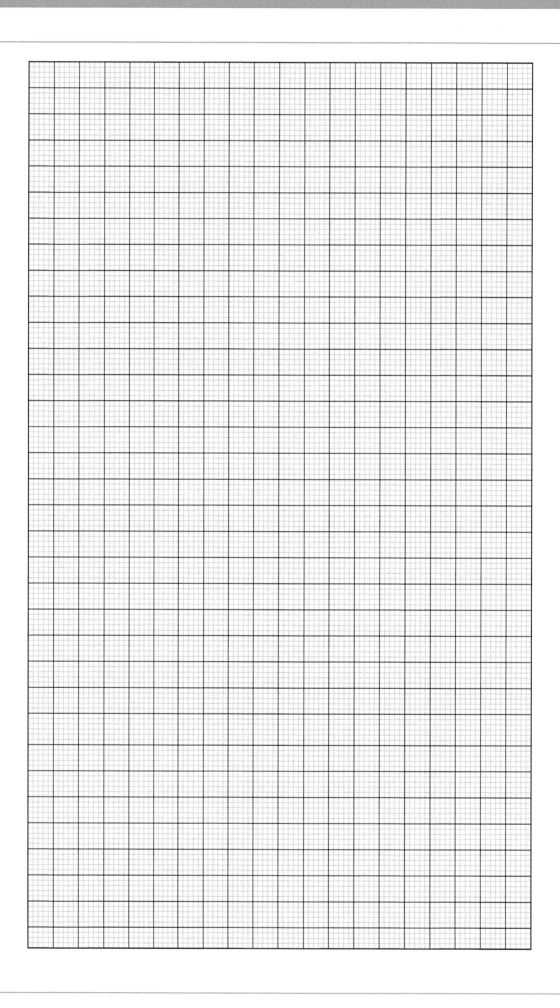

b) Describe and explain the correlation identified. (8 marks)

_____

_____

_____

_____

_____

_____

_____

_____

_____

_____

_____

_____

_____

_____

_____

_____

c) Investigate and explain any residual(s) identified. (3 marks)

_____

_____

_____

_____

_____

_____

_____

		Mark/20
Strengths of the answer		
Ways to improve the answer		

You need to know:

- how to construct and interpret triangular graphs.

Why triangular?

Triangular graphs are used to show data that can be divided into three parts. The three most common applications in geography are studies of:

- soil texture (sand, silt and clay)
- occupational structure (primary, secondary and tertiary employment)
- population structure (young, adult and aged).

The data must be in percentages and the total must add up to 100%. Plotting requires care, but only requires two variables. The third is used to check.

Advantages

Key advantages of triangular graphs are that:

- a large number of different data can be plotted on one graph
- groupings are easily recognised
- dominant characteristics can be shown easily
- they provide a fast, reliable way of showing changes over time (as position on the graph changes).

Sample	Sand	Silt	Clay
A (clay loam)	25	45	30
B (clay)	5	20	75
C (sand)	90	5	5

Figure 1 *Comparative soil textures*

- Triangular graphs are used to show data that can be divided into three components.
- Common applications in geography include showing soil texture, occupational (employment) structure and population structure.

- Key advantages of triangular graphs include the large amount of data, groupings and dominant characteristics that are easily visualised, and they can provide a fast, reliable way of showing changes over time.

Now practise...

Question 1

As a country's economy matures, the proportion of people employed in primary (agriculture, fishing, mining, etc), secondary (manufacturing) and tertiary (services) occupations changes. This is known as occupational (or employment) structure.

Tip

Plotting on a triangular graph
- Use a ruler to guide your eye.
- Always read the obtuse angle.
- Plot with two values – check with the third.

	Primary %		Secondary %		Tertiary %	
	1980 (est)	2015 (est)	1980 (est)	2015 (est)	1980 (est)	2015 (est)
Brazil	30	10	24	40	46	50

🔺 **Figure 2** *Brazil's occupational structure (1980–2015)*

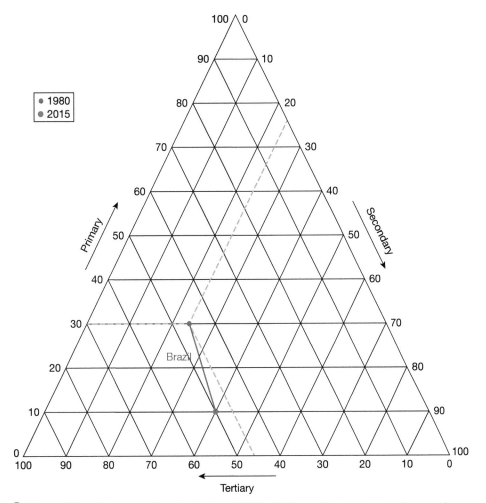

🔺 **Figure 3** *Brazil's occupational structure (1980–2015) plotted on a triangular graph*

Study Figure **4** which shows estimates of the proportion of people employed in primary, secondary and tertiary occupations for six countries in both 1980 and 2015.

	Primary %		Secondary %		Tertiary %	
	1980 (est)	2015 (est)	1980 (est)	2015 (est)	1980 (est)	2015 (est)
Brazil	30	10	24	40	46	50
Burundi	84	94	5	2	11	4
Ethiopia	80	73	7	7	13	20
South Korea	34	5	29	24	37	71
UK	2	1	37	15	61	84
USA	2	1	32	20	66	79

🔺 *Figure 4 Occupational structure for six countries (1980–2015)*

a) On the triangular graph provided, plot the changes in occupational structure (1980–2015) for the six countries in Figure **4**.

(6 marks)

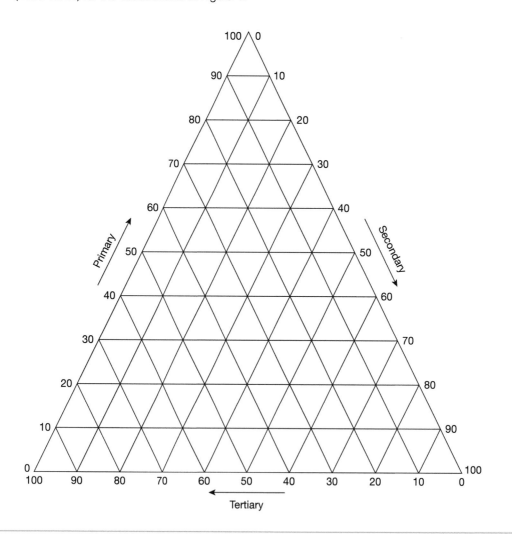

b) State any obvious developmental groupings. **(2 marks)**

_____

_____

_____

c) Describe and explain the resulting patterns with reference to the development process. **(9 marks)**

_____

_____

_____

_____

_____

_____

_____

_____

_____

_____

_____

_____

_____

_____

_____

_____

_____

d) Investigate and explain any anomaly(s). **(3 marks)**

_____

_____

_____

_____

		Mark/20
Strengths of the answer		
Ways to improve the answer		

You need to know:

- when and how to plot radial charts.

Radial charts

Look at Figure **1**. Rather like triangular graphs (see 4.4), radial charts have great value in showing patterns and weightings, with the added potential of allowing comparisons over time.

However, their greatest value is in comparing data showing direction, orientation or alignment. For example, familiar applications are likely to include the:

- direction of winds in weather studies
- orientation of upland corries in glacial studies
- long-axis alignment of glacial till deposits (Figure **2**).

Radial charts tend to be named differently in different circumstances. Unfortunately you can expect no consistency. Most common terms are 'star' and 'rose' charts/diagrams/graphs, but you may also come across terms such as 'radar chart', 'compass rose' and 'polar graph'.

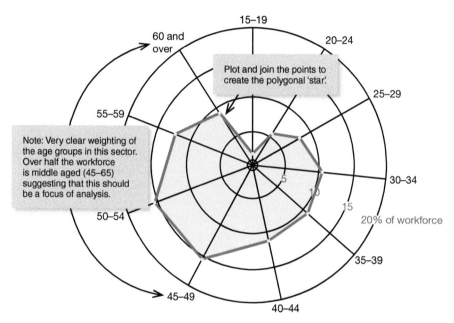

Plot and join the points to create the polygonal 'star'.

Note: Very clear weighting of the age groups in this sector. Over half the workforce is middle aged (45–65) suggesting that this should be a focus of analysis.

Age	% of workforce
15–19	0.86
20–24	5.30
25–29	7.90
30–34	10.60
35–39	10.90

Age	% of workforce
40–44	11.24
45–49	15.35
50–54	15.50
55–59	12.53
60 and over	9.81

▲ **Figure 1** *Age profile of Lincolnshire County Council workforce, March 2017*

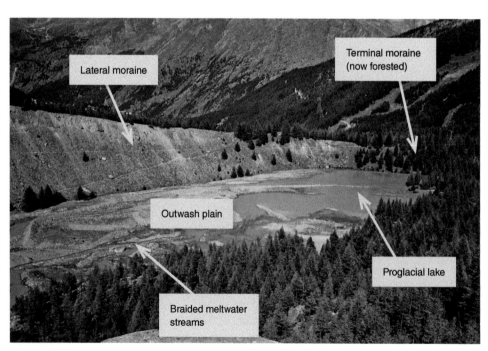

◀ **Figure 2** *Spielboden Glacier outwash plain and moraines – Saas Fee, Valais, Switzerland*

Some might argue that the name applied to a chart depends more on the outcome than the technique. For example, star diagrams tend to be named as such in purely illustrative situations (Figure **1**), where direction is less important – such as a radial chart showing distance travelled to different locations including workplace, main shopping centre, leisure activities and so on.

In contrast, rose charts/graphs are more likely to show clearly quantified, comparative, directional data – plotted using, for example, lines, bars or wedges. In practice, however, adopt whatever term and technique the question demands. They are all radial charts.

Wind rose

One aspect of weather studies is variations in wind direction. Having recorded the wind direction over a period of time it is necessary to process and record the data – most commonly using a wind rose (Figure **3**). From this chart the most usual direction of wind (the prevailing wind) can be determined.

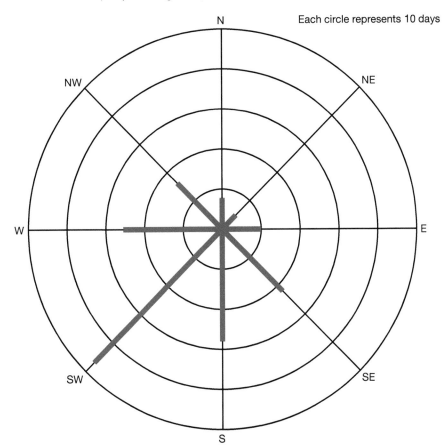

Each circle represents 10 days

Direction	Number of days
N	8
NE	5
E	10
SE	22
S	28
SW	47
W	26
NW	17

 Figure 3 *Measurable wind directions over a six-month period*

Sixty second summary

- Radial charts have great value whenever comparative data (showing direction, orientation or alignment) needs to be illustrated for subsequent analysis.
- Radial charts have different forms and names.
- Points may be plotted and joined on a radial chart to create a polygonal 'star'.
- Lines, bars or wedges may be plotted to create a 'rose'.
- Common applications include demonstrating wind directions in weather studies, corrie orientation in glacial studies and long-axis alignment of glacial till deposits (till fabric analysis).

Now practise...

Question 1

Study Figure **4**.

Bearing (°)	Number of stones	Bearing (°)	Number of stones	Bearing (°)	Number of stones
0	2	120	4	240	8
15	4	135	5	255	5
30	10	150	3	270	3
45	12	165	3	285	3
60	8	180	2	300	4
75	5	195	4	315	5
90	3	210	10	330	3
105	3	225	12	345	3

⬤ **Figure 4** *The orientation of the long-axes of 124 stones in a sample of glacial till*

a) Graph the data on the radial chart below.

(6 marks)

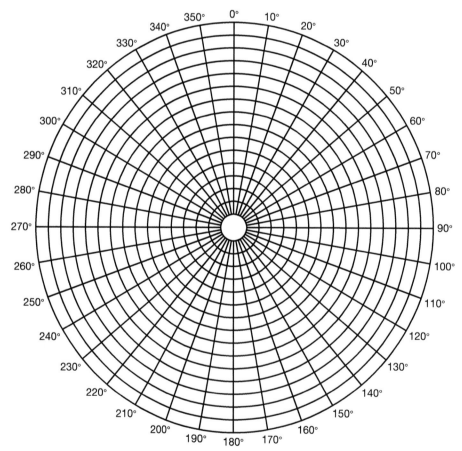

⬤ *Figure 5*

b) Justify the technique adopted. **(2 marks)**

_____

_____

c) From the evidence in the completed graph, suggest the likely direction of ice flow. **(1 mark)**

_____

Extension questions (for candidates studying the _Glacial systems and landscapes_ option)

d) Explain **two** other ways of inferring the direction of ice movement from evidence obtained from glacial deposits. **(4 marks)**

_____

_____

_____

_____

_____

_____

_____

e) Study Figure **2**.

What field evidence in this location would enable you to distinguish between glacial deposits and fluvial deposits? **(7 marks)**

_____

_____

_____

_____

_____

_____

_____

_____

_____

_____

_____

_____

		Mark/9 or 20
Strengths of the answer		
Ways to improve the answer		

- how to represent a large range of data on a single graph
- how and when to use cumulative scales.

Graphical skills

We have already established that there are a great variety of graphs and diagrams that you need to be able to draw and interpret (Figure **1**).

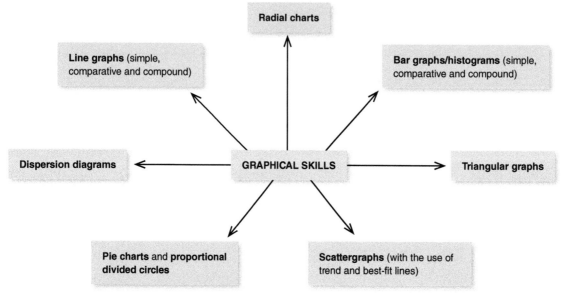

● **Figure 1** *Different types of graphs and diagrams*

In consequence, graph paper comes in many forms and varieties.

But what do you use when you need to represent a large range of data on a single graph (e.g. GDP of countries) or need to compare values that vary enormously in size (e.g. river velocities and sediment sizes)?

The problem with standard graph paper is the inevitable compression required to fit wide data sets – this can hide important variation. Using a broken scale may address this on occasions (see Figure **7**, 4.2), but using logarithmic graph paper solves the problem.

Logarithmic graph paper

There are two types of logarithmic graph paper:

- **Log-log graph paper** has logarithmic cycles along both axes. Each cycle increases by a power of 10 and you can choose the starting point. For example, the first cycle might be from 1 to 10, the second to 100, the third to 1000 and so on. The huge potential range of values possible is represented particularly well in the Hjulström Curve (Figure **2**). On this iconic graph, the relationship between imperceptible river motion to raging torrents, and microscopic particles to boulders the size of cars, can be shown for subsequent analysis.
- **Semi-log graph paper** has arithmetic values along the *x*-axis (the independent variable) increasing by a standard value each time (e.g. 2, 4, 6, 8, 10 and so on). This is useful whenever there is a need to show regular progressions, such as time or distance. However, the *y*-axis (the dependent variable) is logarithmic allowing a wide-ranging scale with greater space for the smaller values (Figure **3**).

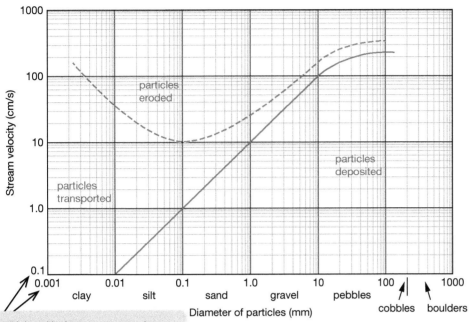

Note: With logarithmic paper you can choose the starting value of each cycle (providing thereafter the cycles progress in sequence).

◀ **Figure 2** *Using log-log paper – the Hjulström Curve*

Note: the logarithmic *y*-axis compresses the vertical scale. A standard arithmetic scale here would require an exceptionally tall graph.

Site	Altitude above sea level (m)	Gradient (°)	Distance from source (km)	Average velocity (m/s)
1	160	3.3	1.0	0.084
2	101	1.0	3.5	0.230
3	54	1.7	8.0	0.290
4	20	0.6	13.0	0.273

🔼 **Figure 3** *Using semi-log paper – showing how average river velocity changes with distance from the source*

Note: The *x*-axis showing 'regular' data – in this case, distance from the river source.

Cumulative scales

Not to be confused with 'cumulative scaling' (Guttman scaling) used in social psychology, cumulative scales in geography are most commonly associated with percentage values progressing to the total of 100%. You are most likely to come across their use when studying inequalities in distribution, such as in population, industry or wealth distributed across a given area.

Plotting graphs with cumulative scales tend to provoke anxiety and errors that are easily avoided in practice. Apply the logic of lap times in a race accumulating towards a total race time, and double-check the accumulation(s) before plotting, and you will find it very straightforward.

Plotting a Lorenz curve is your most likely application (see over).

The Lorenz curve

The Lorenz curve was developed to show and measure any inequality of distribution in graphical form (Figure **4**). It is an excellent technique for showing changes in distribution over time.

Plotting the Lorenz curve is straightforward providing that you:

- rank the data (lowest to highest or vice versa) – in Figure **4**, the proportion of wealth distributed across the UK (2012–14) is ranked from poorest to wealthiest
- double-check that the cumulative totals build to 100%.

The nearer the curve is to the 45° (diagonal) line of perfect equality, the less inequality within that population.

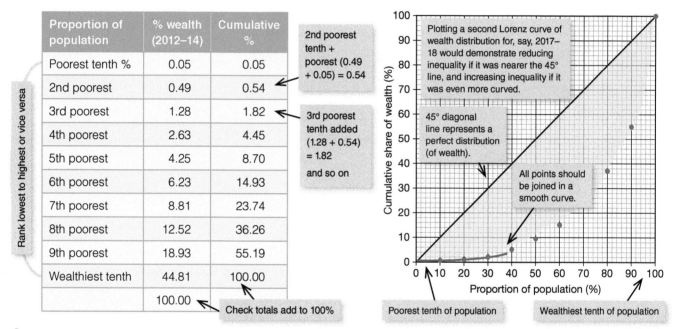

Proportion of population	% wealth (2012–14)	Cumulative %
Poorest tenth %	0.05	0.05
2nd poorest	0.49	0.54
3rd poorest	1.28	1.82
4th poorest	2.63	4.45
5th poorest	4.25	8.70
6th poorest	6.23	14.93
7th poorest	8.81	23.74
8th poorest	12.52	36.26
9th poorest	18.93	55.19
Wealthiest tenth	44.81	100.00
	100.00	

Rank lowest to highest or vice versa

2nd poorest tenth + poorest (0.49 + 0.05) = 0.54

3rd poorest tenth added (1.28 + 0.54) = 1.82 and so on

Check totals add to 100%

Plotting a second Lorenz curve of wealth distribution for, say, 2017–18 would demonstrate reducing inequality if it was nearer the 45° line, and increasing inequality if it was even more curved.

45° diagonal line represents a perfect distribution (of wealth).

All points should be joined in a smooth curve.

Poorest tenth of population

Wealthiest tenth of population

▲ Figure 4 *Plotting a Lorenz curve for the distribution of wealth in the UK (2012–14)*

Measuring inequality – the Gini index

Many would argue that measuring inequality is one of the fundamentals of geography. The Gini index does this by measuring from Lorenz curves to allow:

- comparison of inequality within countries (e.g. comparing rural and urban areas)
- comparison of inequality between countries
- analysis of changing trends through time.

The index measures the area between the Lorenz Curve and the 45° line of perfect equality, expressed as a percentage of the maximum area under the line. (On graph paper this is done by counting the small graph squares.) A value of 0 represents no inequality and a value of 100 corresponds to inequality in its most extreme form – in reality, a single individual having all the wealth in a country!

Sixty second summary

- Graph paper comes in many forms and varieties – including standard, triangular and radial.
- The problem with plotting on standard graph paper, with arithmetic scale progression (e.g. 10, 20, 30, 40, etc), is the inevitable compression required to fit wide data sets – hiding important variation.
- Log paper solves this problem by using scales that progress in logarithmic cycles, increasing by a power of 10 (e.g. 1 to 10, 10 to 100, 100 to 1000 and so on) – you choose the starting point.
- Cumulative scales are most commonly associated with percentage values progressing to the total (100%).
- The Lorenz curve was developed to show and measure any inequality of distribution in graphical form.
- The Gini index measures inequality from Lorenz curves allowing comparison of values and analysis of changing trends.

Now practise...

Question 1

Study Figure **6**.

Continent (ranked in order of population)	Population (est. 2016)	Population (%)	Cumulative population (%)	Area (%)	Cumulative area (%)
Asia	4 436 224 000	59.68	59.68	20.3	
Africa	1 216 130 000	16.36		22.3	
Europe	738 849 000	9.95		20.1	
N. America	579 024 000	7.79		15.8	
S. America	422 535 000	5.68		15.2	
Oceania	39 901 000	0.54	100.00	6.3	
Total	**7 432 663 000**	**100.00**		**100.0**	

🔺 **Figure 6** *Population data by continent*

a) Complete the table. **(2 marks)**

b) Construct a Lorenz curve by plotting cumulative population (%) against cumulative area (%). **(3 marks)**

c) Explain your graph. **(4 marks)**

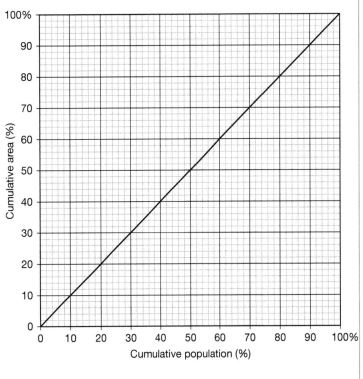

Strengths of the answer	**Mark/9**
Ways to improve the answer	

You need to know:

- that measures of central tendency are concerned with the 'average' of a set of data
- the numerous ways of calculating an average – each of which is imprecise and has a different interpretation
- that not all measures of central tendency are applicable to all types of data.

Measures of central tendency

There are three measures to show the central point of a data set – **mean**, **median** and **mode**.

Mean

This is often referred to as the 'average' and is calculated by adding up the individual values for a set of data and dividing by the number of values. This 'arithmetic mean' has useful applications in geography, such as in calculating the average temperature of a place in climate studies (see 4.2).

But the arithmetic mean can be influenced by outliers or anomalies (e.g. extreme values such as unusually wet or dry years – hence the convention of adopting 30-year averages in plotting climate graphs). The arithmetic mean also gives no indication of how data within a set are spread around the average.

Median

This is the central or mid-point value in a ranked data set. Half of the data set lies above the median and half below. Both ranking and spread are shown using a **dispersion diagram** (Figure **1**). If there is an odd number of values (e.g. 25), the median value is the middle value of the ranking and will be a whole number (i.e. the 13th). When there is an even number of values (e.g. 24), the median is the average of the two middle values (i.e. 12th and 13th).

Mode

This 'modal value' is the value that appears most frequently in a data set. However, as a measure of central tendency it is only really valid when there is a substantial data set – otherwise it can be misleading.

Dispersion, range and inter-quartile range

Look again at the dispersion diagram (Figure **1**).

Dispersion

This refers to how data are distributed within the **range**. Dispersion diagrams are used to show how far data are dispersed or clustered.

Range

This describes the span of data across a set, calculated simply by subtracting the lowest value from the highest value. The range, therefore, describes the spread of the data. However, it gives no idea about how the data are distributed (whether spaced evenly or clustered) and is prone to be affected by outliers lying well beyond the range of the majority.

Inter-quartile range

This is a statistical value to show where the middle 50% of the data lie within any set. It is based around the median, and takes all data into account (discounting the affect of any anomalies). The inter-quartile range can be used to compare dispersions between two or more sets of data.

Calculating the inter-quartile range

To calculate the inter-quartile range you need to divide the ranked data set into four equal quarters.

Start by finding the median value, then the 25% and 75% quartiles. To find the quartiles, count the number of data values either side of the median, e.g. 10 on either side of the median for data set **B** in Figure **1** – and then determine the median of each half.

The median of the upper half is called the upper quartile, and the median of the lower half is called the lower quartile. The difference between the upper and lower quartiles is called the inter-quartile range. (Note that the top of the inter-quartile range is marked by the upper quartile line and the bottom is marked by the lower quartile line.)

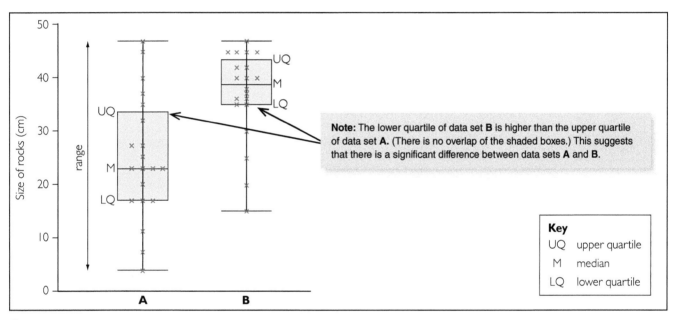

Note: The lower quartile of data set **B** is higher than the upper quartile of data set **A**. (There is no overlap of the shaded boxes.) This suggests that there is a significant difference between data sets **A** and **B**.

Key
UQ   upper quartile
M   median
LQ   lower quartile

Figure 1 *A dispersion diagram showing the median and inter-quartile range (a box and whisker plot)*

 Sixty second summary

- Mean, median and mode all show the central point of a data set – but provide no information about the range or spread of data.
- The mean is the average calculated by adding up the individual values for a set of data and dividing it by the number of values.
- The median represents the central or mid-point value in a ranked data set, so half of the data set lies above the median and half below.

- The mode is the value that appears most frequently in a data set.
- Dispersion diagrams show how data are dispersed or clustered within a range – with the range describing the spread.
- The inter-quartile range (showing where the middle 50% of the data lie within any set) can be used to compare dispersions between two or more sets of data.

Now practise...

Question 1

An exasperated Year 13 candidate in Sheffield is tired of his parents and Scottish grandparents in Nairn bickering as to which place is wettest. He searches online for Met Office annual precipitation data in order to demonstrate whether or not there are significant rainfall differences.

Knowing that his relatives will not appreciate a lecture based upon tabulated raw data, or line graphs that look like oscilloscope traces of erratic heartbeats, he decides to revise and practise measures of central tendency, dispersion, range and inter-quartile range as a means of presenting his results.

Study Figure **2**, which shows annual precipitation totals for both Sheffield and Nairn, 1992–2011.

Year	Annual precipitation total (mm): Sheffield, England	Annual precipitation total (mm): Nairn, Scotland	Year	Annual precipitation total (mm): Sheffield, England	Annual precipitation total (mm): Nairn, Scotland
1992	1345.0	1196.6	2002	1397.0	1241.1
1993	1288.4	1025.3	2003	1661.8	1550.4
1994	1364.6	1390.2	2004	1505.4	1248.6
1995	1586.3	1439.8	2005	1551.6	1251.7
1996	1441.2	1377.8	2006	1605.8	1274.9
1997	1586.4	1264.9	2007	1588.6	1204.3
1998	1396.7	1199.6	2008	1512.5	1178.4
1999	1625.7	1259.5	2009	1595.8	1467.5
2000	1405.1	1307.5	2010	1491.3	1281.0
2001	1422.5	1153.4	2011	1598.9	1345.9

🔼 **Figure 2** *Annual precipitation totals for both Sheffield and Nairn, 1992–2011*

a) On the graph paper opposite, present the data for **both** Sheffield and Nairn on a dispersion diagram.　　(9 marks)

b) For **both** Sheffield and Nairn state:

(i) the mean annual precipitation total _____　(2 marks)

(ii) the median annual precipitation total _____　(2 marks)

(iii) the precipitation range. _____　(2 marks)

c) Refer again to the introduction and suggest how the candidate might explain these results to his bickering relations.　　(5 marks)

_____

_____

_____

_____

_____

_____

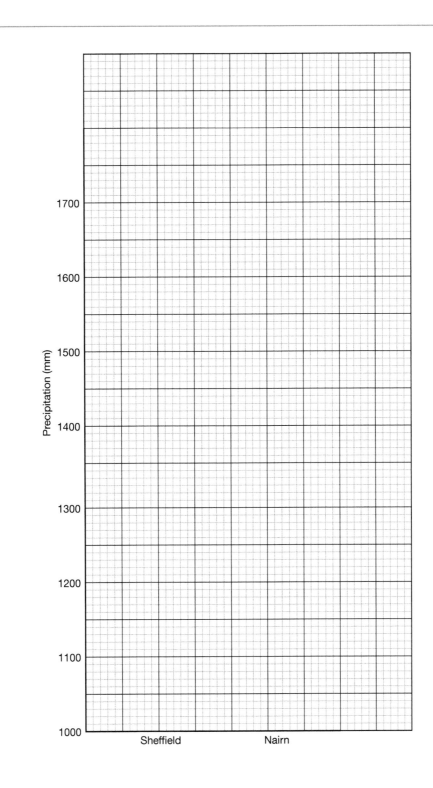

Strengths of the answer		Mark/20
Ways to improve the answer		

Variance and standard deviation

You need to know:

- how to identify and make sense of how data vary
- how to measure the spread of values around the mean.

Data variance

Data variance refers to how data vary. As with the inter-quartile range (see 5.1) high variance means that the data set has a large **range** of values. Low variance would result from data grouped closely around the mean. In effect, data variance shows *deviation* from an average.

But we can illuminate this further by calculating **standard deviation** which measures *by how much* the data varies from the mean. Inter-quartile 'box and whisker plots' (Figure **1**, 5.1) show variation, but standard deviation quantifies it.

By accounting for *all* values in a data set, any explanation adopting these summarising statistics has great use and value when it comes to investigation of *why* there is variation. For example, in a study of glacial outwash (see 4.5), you might want to investigate the reasons for variation demonstrated in material size and/or shape with distance from the glacier snout.

Example: Calculating standard deviation

Year	Rainfall in mm (x)	Variance from mean (x − x̄)	Variance from mean squared (x − x̄)²
2003	680	−177.9	31648.41
2011	722	−135.9	18468.81
2010	732	−125.9	15850.81
2005	750	−107.9	11642.41
2013	811	−46.9	2199.61
2016	824	−33.9	1149.21
2006	840	−17.9	320.41
2017	845	−12.9	166.41
2009	865	7.1	50.41
2015	873	15.1	228.01
2004	895	37.1	1376.41
2007	937	79.1	6256.81
2008	975	117.1	13712.41
2014	990	132.1	17450.41
2012	1130	272.1	74038.41
Total (Σ x) = 12869		Σ (x − x̄)² = 194558.95	
Mean (x̄) = 857.9			

⬥ **Figure 1** *Ranked average total rainfall values for England, 2003–2017*

$$\sigma = \sqrt{\frac{\Sigma(x-\bar{x})^2}{n}}$$

where:

σ = standard deviation Σ = sum of

x = individual value

x̄ = mean   n = number in the sample

$$\sigma = \sqrt{\frac{\Sigma(x-\bar{x})^2}{n}}$$

$$\sigma = \sqrt{\frac{194558.95}{15}}$$

$$\sigma = \sqrt{12979.97}$$

$$\sigma = 113.88$$

This means that the amount by which rainfall in England is likely to deviate either side of the mean is 113.88mm.

The years to which this applies are shown in **bold** on Figure **1**, i.e. in 9 of 15 years (60%), and in Figure **2** as $\bar{x}+1\sigma$ / $\bar{x}-1\sigma$.

In most data sets, twice the standard deviation (in this case 227.76) will normally encompass 95% of data (between 1085.66 and 744.02). In this case, four values lie outside this range (27%), including three less than $\bar{x}-2\sigma$, suggesting an abnormal distribution with a slight positive skew*.

The fifteen years investigated have a median value (845mm), slightly less than the mean (857.9mm), and so more rainfall than 'normal'.

*If the bulk of the values are less than the mean, the distribution is said to be *positively skewed*.

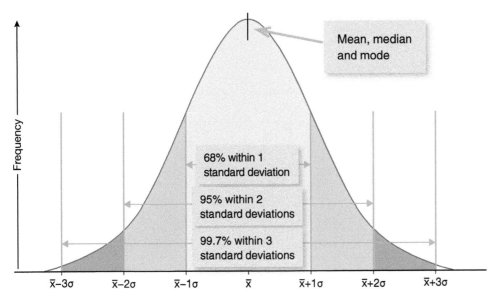

Figure 2 *A normal distribution curve*

Labels on figure:
- Mean, median and mode
- 68% within 1 standard deviation
- 95% within 2 standard deviations
- 99.7% within 3 standard deviations
- Frequency (y-axis)
- $\bar{x}-3\sigma$ $\bar{x}-2\sigma$ $\bar{x}-1\sigma$ $\bar{x}$ $\bar{x}+1\sigma$ $\bar{x}+2\sigma$ $\bar{x}+3\sigma$

Sixty second summary

- Data variance shows how data vary from the mean.
- Standard deviation measures by how much the data varies from the mean.
- Inter-quartile 'box and whisker plots' also show variation, but standard deviation quantifies it.

- Variance and standard deviation account for all values in a data set.
- A normal distribution curve is shaped like a bell, and symmetrical around its mean, medial and modal class.

Now practise...

Having established that rainfall in England from 2003–2017 did not follow the normal, expected distribution, it might be reasonable to expect that Scotland will demonstrate similar 'skewness'. If Scotland's average total rainfall values over the same time period are similarly abnormal, might we consider whether or not both countries demonstrate sufficient evidence to indicate climate change?

Study Figure **3**.

Year	Rainfall in mm (x)	Variance from mean (x − x̄)	Variance from mean squared (x − x̄)²
2003	1205	−394.1	155 314.81
2010	1250	−349.1	121 870.81
2013	1465	−134.1	17 982.81
2016	1500	−99.1	9 820.81
2017	1504	−95.1	9 044.01
2007	1590	−9.1	
2005	1596	−3.1	
2012	1609	9.9	
2006	1651	51.9	
2009	1694	94.9	
2004	1700		
2008	1729		
2014	1752		
2015	1851		
2011	1890		
Total (Σx) = 23 986		Σ(x − x̄)² =	
Mean (x̄) =			

▲ **Figure 3** *Ranked average total rainfall values for Scotland, 2003–2017*

a) Complete the table (including calculating the mean). (3 marks)

b) State the median value. (1 mark)

___

c) Calculate the standard deviation using the formula:

$$\sigma = \sqrt{\frac{\Sigma(x-\bar{x})^2}{n}}$$

where: σ = standard deviation Σ = sum of
x = individual value
x̄ = mean n = number in the sample

Show your working in the space below. (4 marks)

d) (i) On Figure **3** highlight the years within 1σ either side of the mean. **(2 marks)**

(ii) State how many years this is, and as a percentage of the total. **(1 mark)**

_____

_____

e) In most data sets, twice the standard deviation will normally encompass 95% of data. Does Scotland's average total rainfall over the fifteen-year period demonstrate a normal distribution? (Justify your answer.) **(3 marks)**

_____

_____

_____

_____

_____

_____

f) Does the data presented in the tables in Figures **1** and **3** provide sufficient evidence to prove climate change? (Justify your answer.) **(6 marks)**

_____

_____

_____

_____

_____

_____

_____

_____

_____

_____

_____

_____

		Mark/20
Strengths of the answer		
Ways to improve the answer		

You need to know:

- why, when and how to use the Spearman rank correlation coefficient test.

Testing the strength of correlation

Correlation describes the degree of association between two sets of data. Scattergraphs (4.3) are ideal for visualising such an association, following the convention of plotting the independent variable on the *x*-axis and the dependent variable on the *y*-axis. The degree of correlation (e.g. positive, negative, strong, poor or absent) is then estimated by the closeness of the plotted points to the best-fit line, drawn by eye and showing any trend. Residuals (outliers) lying well beyond the best-fit line are ignored as anomalous, but inevitably prove to be illuminating in analysis of the relationship that follows.

However, in some instances the arrangement of the points on a scattergraph makes it impossible to draw a line of best-fit, inferring that there is no correlation. Consequently, statistical testing to see if there is any correlation and, better still, to quantify the strength of it would be invaluable.

The *product moment correlation* has its use in this respect, but is not straightforward and restricted by the assumption of normally distributed variables. In contrast, the *Spearman rank correlation coefficient* test is simple to calculate and allows you to state the degree of confidence that the relationship has not occurred by chance.

Example: Is there a correlation between soil moisture content and the number of alders?

Site no.	Soil moisture % (x)	Rank (r_x)	No. of alders (y)	Rank (r_y)	d	d²
1	21	8	47	5	3	9
2	18	9	35	8	1	1
3	9	10	28	9	1	1
4	25	7	27	10	−3	9
5	27	6	44	7	−1	1
6	38	3	66	3.5	−0.5	0.25
7	33	4	66	3.5	0.5	0.25
8	48	2	75	1	1	1
9	49	1	73	2	−1	1
10	32	5	45	6	−1	1
n = 10						Σd^2 = 24.5

n = number of data pairs (in this case 10)

d = difference between the ranks of each of the paired variables

Σ = sum of the values (in this case the sum of all the d² values)

R(s) = rank correlation coefficient. This ranges between −1.0 (perfect negative correlation) and +1.0 (perfect positive correlation). A value of 0.00 indicates no correlation.

Note: Where two or more sites have the same values (e.g. the number of alders at sites 6 and 7) calculate an average rank value. (Hence $\frac{3+4}{2}$ = 3.5.)

🔺 **Figure 1** *Correlation between soil moisture content and the number of alders*

$$R(s) = 1 - \frac{6\Sigma d^2}{n^3 - n}$$

$$= 1 - \frac{6 \times 24.5}{1000 - 10}$$

$$= 1 - \frac{147}{990}$$

$$= 1 - 0.15$$

$$= 0.85$$

Testing for significance

For greater precision, any correlation coefficient should be tested for **significance**. Significance is a mathematical term, which allows the user to get a measure of confidence in the correlation they have calculated.

In geography a 95% confidence level is adequate. 99% confidence is, however, desirable because we can state with certainty that there is only a 1% chance of error.

	Significance (confidence) level	
n	0.05 (95%)	0.01 (99%)
8	0.643	0.833
9	0.600	0.783
10	0.564	0.745
11	0.536	0.709

Figure 3

In the example of correlation between soil moisture content and number of alders, R(s) = 0.85. So we can state with 99% certainty that there is a positive correlation between soil moisture content and the number of alders. (The null hypothesis of no relationship can be rejected.)

However, a statistical acceptance is only as good as the geography! You have to judge if, geographically, it makes sense for a causal relationship to exist, i.e. one variable (soil moisture content) influencing the number of alders. In short, further investigation can now determine the cause of this relationship.

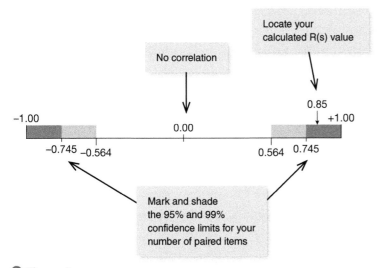

Figure 4

Figure 2 Alder woodland

Tip

Anticipate and visualise the strength of the correlation
Simple arithmetic errors (such as misreading and/or misplacing a negative in the final section of the calculation) and misreading confidence levels are common. Yet these are simple to check by both anticipating and visualising the strength of the correlation (Figure 4).

By sketching your own confidence line for the relevant number of paired variables you can:
- check that your value lies between −1.00 (perfect negative correlation) and +1.00 (perfect positive correlation)
- visualise the strength of the relationship at a glance
- confirm it makes sense (are you expecting a correlation to be absent, strong or weak, positive or negative?).

Sixty second summary

- Correlation describes the degree of association between two sets of data.
- The *Spearman rank correlation coefficient* is a statistical test which shows the strength of a relationship between two sets of data.
- Ten sets of data (such as systematic sampling sites) is the minimum number to be used to provide a sufficient spread of data.
- Even if the test shows a significant correlation it does not prove that there is necessarily a *causal* relationship between variables.

Now practise...

Question 1

Study the data in Figure **5**.

Country (2017 est.)	GNP per capita (PPP) $ (x)	Rank (r_x)	Consumption of finished steel products (kg) per capita (y)	Rank (r_y)	d	d^2
Egypt	13 000		125			
Turkey	26 500		428			
Argentina	20 700		96			
South Korea	39 400		1130			
Venezuela	12 400		24			
India	7 200		63			
Spain	38 200		273			
Italy	38 000		405			
UK	43 600		164			
Germany	50 200		500			
Japan	42 700		493			
USA	59 500		283			
n =					Σd^2 =	

▲ **Figure 5** *The relationship between Gross National Product per capita (PPP) and consumption of finished steel products (kg) per capita for selected countries in 2017*

   a) Complete Figure **5**.   **(2 marks)**

   b) Calculate the Spearman rank correlation coefficient using the formula:

$$R(s) = 1 - \frac{6\Sigma d^2}{n^3 - n}$$

       where: $R(s)$ = the Spearman rank correlation coefficient

                     n = the number of countries

                     Σd^2 = the sum of differences squared

   Show your working in the space below.   **(2 marks)**

c) State the significance of your result. **(1 mark)**

_____

_____

_____

Significance (confidence) level		
n	0.05 (95%)	0.01 (99%)
10	0.564	0.745
11	0.536	0.709
12	0.503	0.678
13	0.484	0.648

d) Explain the relationship between GNP per capita (PPP) and steel consumption. **(4 marks)**

_____

_____

_____

_____

_____

_____

_____

_____

_____

_____

_____

		Mark/9
Strengths of the answer		
Ways to improve the answer		

Measuring distributions

The **Chi-squared test** is a useful method for testing the difference between grouped data sets – comparing a set of data against how it might be, should everything be distributed evenly. For example, Chi-squared testing could determine how clustered the distribution of marram grass is along a coastal psammosere from the high-water mark inland (Figure **1**).

Statisticians talk of testing 'goodness of fit' whereby the set of frequencies actually observed (O) are tested against a theoretical or 'expected' set of frequencies representing equal chance (E).

$$X^2 = \Sigma \frac{(O-E)^2}{E}$$

where: X^2 = Chi-squared Σ = sum of
O = observed frequencies
E = expected frequencies

▲ **Figure 1** *Marram grass stabilising sand dunes at Studland in Dorset*

The value calculated can then be checked against significance tables in order to see whether or not it is statistically significant. (Demonstrating statistical significance would suggest that the distribution of marram grass is closely linked to distance from the high-water mark.)

Example: Is there a difference in the distribution of marram grass plants with increasing distance from the coastal high-water mark?

This could be stated as a null hypothesis: 'There is no difference in the distribution of marram grass plants with increasing distance from the high-water mark.'

Sampling point	Distance from high-water mark (m)	Number of marram grass plants observed within a plant quadrat (O)	$\frac{(O-E)^2}{E}$ where E = 56/8 = 7
1	10	1	$(1-7)^2/7 = 5.14$
2	20	11	$(11-7)^2/7 = 2.29$
3	30	15	$(15-7)^2/7 = 9.14$
4	40	14	$(14-7)^2/7 = 7.00$
5	50	9	$(9-7)^2/7 = 0.57$
6	60	5	$(5-7)^2/7 = 0.57$
7	70	1	$(1-7)^2/7 = 5.14$
8	80	0	$(0-7)^2/7 = 7.00$
		56	$X^2 = \Sigma \frac{(O-E)^2}{E} = 36.85$

If the null hypothesis was true, the marram grass plants would be distributed equally with 7 plants at each of the eight sampling points (56/8). So 7 is the expected frequency (E).

▲ **Figure 2** *Distribution of marram grass relative to high-water mark*

Testing for significance

$X^2 = 36.85$ means nothing by itself – it has to be checked for statistical significance. For this we use tabulated tables setting out confidence levels against the appropriate *degrees of freedom* (representing the size of the sample minus 1). In this case there are 8 sampling points, so $8 - 1 = 7$ degrees of freedom.

Degrees of freedom	Significance (confidence) level	
	0.05 (95%)	0.01 (99%)
6	12.59	16.81
7	14.07	18.48
8	15.51	20.09
9	17.00	21.67

Checking 36.85 against a standard table showing all levels of confidence for X^2 shows our value exceeding 99.9% confidence. Given that in geography a 95% confidence level is taken as reasonable proof, the conviction that this result has not occurred by chance is incontrovertible.

 Figure 3

36.85 far exceeds both the 14.07 required for 95% confidence and even the 18.48 required for 99% confidence! Consequently, we can reject the null hypothesis with only 1% chance of error. Clearly the distribution of marram grass is closely linked to distance from the high-water mark. Further investigation can now determine the cause(s) of this.

Chi-squared testing is straightforward, providing you take care to double check each calculation. (Showing all stages of your working will help not only you, but also the examiner.)

However, you should be aware of some restrictions:

- The number of observations must exceed 20.
- You cannot use percentages – the data must be numerical.
- The expected frequency for any one group must exceed 5.

Sixty second summary

- The *Chi-squared test* is a useful method for comparing grouped data sets – comparing a set of data against how it might be, should everything be distributed evenly.
- Data must be numerical, but not in percentages.
- The number of observations must exceed 20.
- The expected frequency for any one group must exceed 5.

- For most geographical investigations, significance testing to 95% confidence demonstrates a reasonable likelihood that the result of a statistical test has not occurred by chance. 99% confidence would represent only 1% chance of error.

Now practise...

Question 1

A candidate is investigating a section of beach where granite, limestone and chert (flint) are found in equal quantities. She suspects that because each parent rock has a different vulnerability to weathering and erosion, the proportion of beach pebbles will not be distributed evenly.

Study the data in Figure **4**.

Rock type	Number of pebbles
Granite	186
Limestone	180
Chert	234

▲ *Figure 4 Candidate's random sample of 600 pebbles collected on the beach*

a) State a null hypothesis appropriate for Chi-squared testing of her sample data. **(2 marks)**

b) Is Chi-squared testing for this investigation appropriate, and the size of her sample adequate? **(4 marks)**

c) Calculate the Chi-squared value of her sample using the formula:

$$X^2 = \Sigma \frac{(O - E)^2}{E}$$

where: X^2 = Chi-squared, Σ = sum of, O = observed frequencies, E = expected frequencies

Show your working in the space below. **(4 marks)**

d) State the significance of your result. (1 mark)

Significance (confidence) level		
Degrees of freedom	0.05 (95%)	0.01 (99%)
1	3.84	6.64
2	5.99	9.21
3	7.82	11.35
4	9.49	13.28

_____

_____

_____

_____

e) Suggest a likely explanation for her result. (9 marks)

_____

_____

_____

_____

_____

_____

_____

_____

_____

_____

_____

_____

_____

_____

_____

_____

_____

_____

_____

		Mark/20
Strengths of the answer		
Ways to improve the answer		

Ordnance Survey map symbols

ROADS AND PATHS

M I or A 6(M)	Motorway
A 35	Dual carriageway
A 31(T) or A 35	Trunk or main road
B 3074	Secondary road
	Narrow road with passing places
	Road under construction
	Road generally more than 4 m wide
	Road generally less than 4 m wide
	Other road, drive or track, fenced and unfenced
>>> >	Gradient: steeper than 1 in 5; 1 in 7 to 1 in 5
Ferry	Ferry; Ferry P – passenger only
....	Path

PUBLIC RIGHTS OF WAY

(Not applicable to Scotland)

1:25 000	1:50 000	
		Footpath
		Road used as a public footpath
		Bridleway
		Byway open to all traffic

RAILWAYS

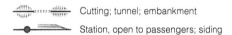

- Multiple track
- Single track
- Narrow gauge/Light rapid transit system
- Road over; road under; level crossing
- Cutting; tunnel; embankment
- Station, open to passengers; siding

BOUNDARIES

- National
- District
- County, Unitary Authority, Metropolitan District or London Borough
- National Park

HEIGHTS/ROCK FEATURES

- Contour lines
- ·144 Spot height to the nearest metre above sea level

outcrop   cliff   scree

ABBREVIATIONS

P	Post office	PC	Public convenience (rural areas)
PH	Public house	TH	Town Hall, Guildhall or equivalent
MS	Milestone	Sch	School
MP	Milepost	Coll	College
CH	Clubhouse	Mus	Museum
CG	Coastguard	Cemy	Cemetery
Fm	Farm		

ANTIQUITIES

VILLA	Roman	✕	*Battlefield* (with date)
Castle	Non-Roman	☆	*Tumulus/Tumuli* (mound over burial place)

LAND FEATURES

- ruin Buildings
- Public building
- Bus or coach station
- Place of worship { with tower / with spire, minaret or dome / without such additions }
- ° Chimney or tower
- Glass structure
- Ⓗ Heliport
- △ Triangulation pillar
- Mast
- Wind pump / wind generator
- Windmill
- Graticule intersection
- Cutting, embankment
- Quarry
- Spoil heap, refuse tip or dump
- Coniferous wood
- Non-coniferous wood
- Mixed wood
- Orchard
- Park or ornamental ground
- Forestry Commission access land
- National Trust – always open
- National Trust, limited access, observe local signs
- National Trust for Scotland

TOURIST INFORMATION

- P Parking
- P&R Park & Ride
- V Visitor centre
- i Information centre
- ☎ Telephone
- ⚕ Camp site/ Caravan site
- Golf course or links
- Viewpoint
- PC Public convenience
- Picnic site
- Pub/s
- Museum
- Castle/fort
- Building of historic interest
- Steam railway
- English Heritage
- Garden
- Nature reserve
- Water activities
- Fishing
- Other tourist feature
- Moorings (free)
- Electric boat charging point
- Recreation/leisure/ sports centre

WATER FEATURES

Marsh or salting   Slopes   Cliff   High water mark
Towpath   Lock   Flat rock   Low water mark
Aqueduct   Ford   Lighthouse (in use)
Canal   Sand   Dunes
Weir   Normal tidal limit   Beacon
Bridge   Lighthouse (disused)
Lake   Footbridge   Mud   Shingle
========= Canal (dry)